Praise for *To Forever Inhabit This Earth*

"This book is a treasure.... Rabbi Nina Beth Cardin's depth of knowledge, and her capacity to pull from a vast catalog of sources—nimbly quoting a sixteenth-century sage on one page and the creator of Calvin and Hobbes on another—brings forth a rich tableau, layered in story and midrash and little-known esoterica.... This belongs on a shelf with the classic tomes—alongside Rachel Carson's and others."

—Barbara Mahany, author of *The Book of Nature: The Astonishing Beauty of God's First Sacred Text*

"Whether you are more familiar with Mary Oliver, Wendell Berry, and Aldo Leopold, or Ben Zoma, Nachman of Bratslav, and Rav Kook, you will be surprised, challenged and inspired by this passionate call to engage in the sacred task of repairing and maintaining a habitable world."

—Dr. Mirele B. Goldsmith, co-founder of Jewish Earth Alliance

"At this time of climate destruction, when it can feel overwhelming to even know where to begin, *To Forever Inhabit This Earth* offers deep Jewish grounding for how to face the climate crisis, firmly rooted in our values."

—Rabbi Jennie Rosenn, founder and CEO of Dayenu: A Jewish Call to Climate Action

"Rabbi Nina Beth Cardin combines history, facts, and moral reasoning with transcendent spirituality and deep wisdom, most notably from her engaging renderings of the 'sacred literary legacy of the Jewish people.' To use one of Cardin's many lovely phrases, this book is itself a 'grand weave.' It is also a beautiful blessing."

—Karenna Gore, executive director of the Center for Earth Ethics at Union Theological Seminary

"Soaringly lyrical and eminently practical, achieving a rare union of body and soul, matter and spirit. Deeply rooted in Jewish sources, it is completely accessible to all looking for an inspiring vision for today's troubled world."

—Dr. Jeremy Benstein, co-founder of the Heschel Center for Sustainability, author of *The Way Into Judaism and the Environment*

"Rabbi Cardin has gifted us with a deeply beautiful, accessible, and relatable framing for understanding and enacting Judaism as guidance toward earthly stewardship as both an inherited responsibility, and a path to Divine meaning in our time."

—Rabbi Andrue Kahn, editor of *The Sacred Earth: Jewish Perspectives on Our Planet*

TO Forever Inhabit THIS Earth

An Ethic of Enoughness

By Rabbi Nina Beth Cardin

BEHRMAN HOUSE
www.behrmanhouse.com

Published by Behrman House, Inc.
Millburn, New Jersey 07041
www.behrmanhouse.com

ISBN 978-1-68115-093-2

Library of Congress Cataloging-in-Publication Data

Names: Cardin, Nina Beth, author.
Title: To forever inhabit this Earth : an ethic of enoughness / by Rabbi Nina Beth Cardin.
Description: Millburn, New Jersey : Behrman House, Inc., [2025] | Includes bibliographical references and index. | Summary: "Rabbi and environmental activist Nina Beth Cardin draws upon Jewish texts in this call to action as the climate crisis persists"— Provided by publisher.
Identifiers: LCCN 2024030555 (print) | LCCN 2024030556 (ebook) | ISBN 9781681150932 | ISBN 9781681151731 (ebook)
Subjects: LCSH: Human ecology—Religious aspects—Judaism. | Sustainability—Religious aspects—Judaism. | Environmental ethics.
Classification: LCC BT695.5 .C3628 2025 (print) | LCC BT695.5 (ebook) | DDC 296.3/8—dc23/eng/20240806
LC record available at https://lccn.loc.gov/2024030555
LC ebook record available at https://lccn.loc.gov/2024030556

Design by Harriet R. Goren
Edited by Rabbi Deborah Bodin Cohen
Editorial Consultants: Rabbi Daniel Swartz

The publisher gratefully acknowledges the following sources of images:
COVER: Courtesy of Shutterstock: Leka Sergeeva
INTERIOR: Courtesy of Shutterstock: Katyau

♻ Printed on recycled paper.
Printed in China
9 8 7 6 5 4 3 2 1

For Avram

צורי אחסה בו

Contents

Prologue

I like my desk. It is a repurposed, mid-century rolling beverage cart, the kind that beehive-coiffed suburban hostesses would have steered effortlessly around their living rooms, eliminating the pesky distance between guest and bar. I sit at the end of one of its extended side panels and imagine the conversations that swirled around it, everyone celebrating those heady postwar days when life was good, gas was cheap, the future was bright, and anything seemed possible.

And yet, as I open my laptop to begin this book, my desk crowds in on me. A riotous congestion of things have congregated there: a coffee mug, a pool of library books, piles of papers—read and unread—my cell phone, a cache of paper clips, errant toys left by visiting grandchildren.

"I have no space to think," I fuss. Out loud. Here I am, writing about enoughness and sustainability, yet burdened by the press of my own possessions.

I regret to report that my first impulse to this dilemma was to open wide my arms, call the world as my witness, and righteously proclaim, "I need a bigger desk!" Not: I need to get better organized. Or: I need to clean up. Or: It's time to declutter. Or (let the truth be told): I'm stuck on how to start the book. Instead, my first impulse was: I need a bigger desk.

And in that impulse lies the problem.

Now, in my defense, my desk is modest by modern standards, just 30" x 30" (30" x 40" when fully extended). So while it is quite serviceable, it is not large. Still, it is significantly larger than many antique desks (30" x 24") and mammoth when compared with ladies' writing desks (26" x 18"), on which whole worlds of heartaches, consolations, forbidden loves, undying friendships were played out.

And as if that weren't enough to chasten me, Emily Dickinson's desk was only 17 ⅜ inches square.[1]

The irony of the moment did not go unnoticed; here I was, an advocate of sufficiency, an evangelist of enoughness, thinking "more" was the solution to my dilemma rather than less. But it was my misuse of abundance that was the problem, not a suffering of scarcity.

The heady questions of humanity's right relationship with sustainability came alive in the deskscape before me. Questions such as: How much do I—do any of us—need? How much is just right? How much is too much? How much of all we seek is guided by our personal needs and appetites? How much are we beholden to the culture around us? What, in short, is the true measure of enoughness, for any and all of us?

And beyond questions of magnitude loom the questions of methods. How shall we engage with the gifts of the material world? Too often we see nature as a resource, a commodity existing to serve our needs, more than a life force commanding our reverence. We focus on the utility of its matter—what we get from using the stuff of the earth—and not on the sustainability of the process of accessing it, or the awesomeness of its mere presence.

We also often forget that despite its size, the earth is a fragile creation. It has taken four and a half billion turbulent years for the earth to bring forth this world of vibrancy and variety: myriads of stunning, interacting, pulsing life forms, including and especially us. For we humankind, latecomers though we are to this dynamic party, are uniquely possessed with the capacity not just to be part of it all but to know—to be aware—that we are part of it all, and to be able to manipulate it for our own benefit.

But such skills and facility have come at a price. The earth is telling us there is something wrong. With it, with us. Because of us. Our numbers, our demands, our processes are disrupting life's equilibrium. Earth is telling us we need to rethink what we are doing and realign our behavior with its cycles and needs. To do that, we need to adopt a new environmental ethic in which we see ourselves not primarily as consumers but as part of the grand weave of nature, sharing the blessings of enoughness with all.

It will take a titanic upheaval, a redirection from the ethic that guided us through recent centuries past, to reimagine our relationship with the earth. But that is what the future—our children, and grandchildren, and their children across the generations—rightfully demands.

How we begin to articulate that new ethic, how we motivate ourselves to live fully into it— as Jews and others—is what this book is all about.

CHAPTER ONE

A Covenant of Companionship

Everything is intended for the establishment
of the world and its continuity,
for the Holy One wanted the world
to be established and endure, as it says:
"God did not create the world to be chaos;
God fashioned it to be habitable."
—*Mishneh Halachot* XIII, 13

A parable: In the beginning, when God began to create the world, God took fire, water, and air and fashioned a magnificent stone out of them. God then cast the stone into the opening of the great abyss, where it lodged to hold back the tumultuous waters of the Deep.

"Know this," the stone told all who sought to remove it, "I must not be lifted, for I hold back the waters of the Deep that would flood the world if left unbounded." But when King David saw the shimmering stone, he desired to know its power and what it was made of. So he ignored the warning and ordered his servants to lift the stone. When his minions dutifully raised the stone, the king peered beneath, into the gaping abyss it had covered. Immediately he heard a loud rushing sound rising toward him and knew he had loosed the raging waters of the Deep.

Quickly seeking a way to undo what he had done, King David turned to his advisors and asked, "Perhaps if we write the name of God on a potsherd and cast it into the depths we might be saved. Who knows if this might work and if it is permitted?"

All his advisors quaked, fearing to answer such a question.

Finally one brave advisor, Ahitophel, stepped forward and dared an answer. "Surely," he offered, "the name of God can be used to bring peace to the world."

So David picked up a potsherd, scratched the name of God onto it, and cast it into the abyss.

Immediately, the waters receded.[2]

That, in short, is what this book is all about: repairing the breach we have caused in our relationship with nature, stemming and reversing the ways the world has been harmed by our mistaken, unwise, and sometimes greedy acts.

We too, like King David, find the world awesome and alluring. We too seek to understand it, explore it, know it better, use it better, plumb its secrets. That indeed is what drives civilizations forward. Yet pursued to excess, without concern for consequences or restraint, such drive also causes us to tread on nature's vulnerabilities and violate its boundaries. For while the earth is large and powerful, it has limits that must be honored. And like the stone in the story, the earth itself warns us when we threaten the systems that keep us all safe. Like Ahitophel, we need to speak up in the name of the greater good. And like King David, we need to respond.

For despite what we have done, despite the harms we have caused, if, like King David, we dare to act boldly (which we can), based on the best guidance of our advisors (which we have), and recruit willing hearts (which we possess), we can heal the wounds, remedy our mistakes, and remake a safe and vibrant world for ourselves, our children, and all the generations to come. It is up to us.

A Jewish Pathway to a Sustainable World

Such a pursuit raises many questions that go to the heart of who and what we are. Technical questions such as: How do we responsibly power a world to meet the demands of our infinite curiosity and expanding needs? How can we best promote regenerative agricultural practices so we nourish the earth as it nourishes us? How do we house and clothe and feed ten billion people while leaving enough world for the rest of Creation to thrive?

And spiritual questions: How much do I—do any of us—really need? How much is just right? How much is too much? In a world in which natural resources are inequitably dispersed, how do we ensure their just and equitable distribution? In a world where humans have such extraordinary technological capacity, how should we define our environmental ethic?

And ethical questions: How do all our things affect the people who make them? The workers who grow, mine, harvest, cut, transport, process, package, and deliver our goods? What about the workers who process the products we have discarded? To live in right relationship with the earth requires that we live in right relationship with one another. Every material thing that fills our daily lives has a story to tell about its impact on the earth and those who brought it into being. And thus, in our use of it, it tells a story about us.

This brings us to an uncomfortable truth: Even good people with good intentions likely behave in ways that harm the earth. Often we have no choice. We cannot recycle if the materials we use are not recyclable. Our discards cannot be given new life if there is no industry to take what we "throw away." We cannot stop destroying our oceans and waterways if our detergents, cos-

While we are succeeding in driving down energy consumption in some sectors, like buildings, transportation, and manufacturing, we are also exponentially increasing our energy needs through the expanded use of smart appliances, cryptocurrency, and especially artificial intelligence and its data centers.

Government-subsidized foods: Having the government assist farmers so they can weather the mercurial nature of growing food is advantageous to all of us. Yet the World Resources Institute reports that of some $600 billion in farm subsidies worldwide, only 5 percent supports sustainable farming.[3] Small farmers, often those using regenerative farming practices, benefit least from these government subsidies.[4] This means that fertilizer- and pesticide-heavy crops are the most subsidized, as is meat. In America, only 0.4 percent of US farm subsidies go to growing fruits and vegetables,[5] which is why it is often cheaper to buy a hamburger than a pepper. Securing the financial health of our worldwide farming industry is no doubt a complicated business. But there is much governments can do to promote equitable and sustainable practices so that earth, harvests, and people all flourish.

metics, soaps, plastics, and fertilizers are made of pollutants. We cannot take public transportation if there is none. We cannot afford the foods that are sustainably raised if unsustainably raised foods are given most of the government subsidies. We cannot choose to own less and share more if there are no collaborative economies to help us.

The pathway to a sustainable world, then, is not just personal. It depends equally on government, science, industry, and the marketplace. So the questions expand: What will it take for society to create the structures we need? What norms and habits need to be dismantled or reclaimed so that our collective appetites and behaviors align with the demands of a healthy world? How do we, in short, embrace the world as "a communion of subjects, not a collection of objects"[6] as the Catholic "ecologian" Thomas Berry challenges us?

These are necessary questions to ask those who run governments and businesses and those of us who possess the blessings of abundance. But what of those of us who suffer from the opposite: the insecurity of scarcity? Billions of people around the world live on the knife's edge, lacking access to clean water, fertile soils, sufficient food, and sustainable economic opportunities.[7] Two and a half billion people live in water-stressed countries; over half a billion people worldwide live in poverty; one-third of the world's population suffers food insecurity. As reported by the United Nations, 3.3 million children in Ethiopia, Kenya, and Somalia (most often girls) are at risk of dropping out of school due to drought. Their families rely on them to walk great distances to fetch the water they need.[8] What do we dare ask of those among us who struggle to merely survive? How can we help build an economy that works for them *and* rebuilds a sustainable earth?[9]

In short, how do we create an equitable, regenerative culture—a culture that contributes to the enduring health of all people and creatures on the earth, both today and tomorrow? Living sustainably should not be an option. It should not be an expression or burden of individual choice. It should be the natural, inevitable practice of a thriving culture we build together.

Our Generation's Great Work: Protecting the Planet

Over the past decades, scientists, economists, environmental activists, public servants, and clergy have been leading the way.[10] They have shown us what we can do that is right, what we have done that is wrong, what needs to be corrected, and as best they can, how to move forward.[11] We know we need to curb greenhouse gas emissions by moving away from fossil fuels and are creating ways to do that. We know we need to transition from our industrial farm system to a method that strengthens both local economies and the soil as we feed the world and are creating ways to do that. We know we create too many plastics and compounds that persist in the environment, clog the life stream, and poison our bodies, and we know we need to work harder to prevent that. We know we need to reimagine how we think about waste and are creating solutions for reducing and recycling what we discard.

But if knowing how to fix the problem were sufficient, if being forewarned had truly enabled us to be forearmed, then we would not be in this mess. For we have known for at least a half century that we were threatening the climate of our planet with our fossil

Collaborative economies: "Why *buy* a drill when all you need is the hole?" That is a question that motivates the thinking of collaborative, or sharing, economies. Why own something, in other words, when you simply need to use it for a specific task or a limited amount of time? We can limit what we need to buy, own, and store if we can share our tools and resources. Collaborative economies can be as vast as home-share industries or as modest as buy-nothing groups in which people in a neighborhood post what they are seeking to get rid of—or what they are seeking to get—and others in the group respond. I have successfully divested myself of sofas, chairs, rocking horses, and more through our buy-nothing group. Today (I just checked), people in my neighborhood are offering everything from a twenty-dollar pet food coupon to a couple of swivel/rocking chairs to a pack of raw hazelnuts!

Regenerative: The United Nations chose the word "sustainable" to define our task of "meeting the needs of the present without compromising the ability of future generations to meet their own needs." While this remains an essential goal, to me the word "sustainable" feels a bit stilted, uninspiring. In recent years, the word "regenerative" has emerged to speak about new (reclaimed) ways of farming, forest management, medicine, and more. At its root, "regenerative" means to create a system that can constantly renew itself, year after year, decade after decade, generation after generation. It points to the future. It points to growth. It speaks of today and tomorrow. It offers hope.

fuels, yet we failed to muster the commitment to change.[12] We know intuitively that nothing can grow forever and ever, and yet we choose to believe that the economy and our consumption can.

The times call for us to reclaim a heightened appreciation of the world in whose graces we dwell. Imagine if we saw the world not as a commodity but as a fragile, improbable, extraordinary gift born of billions of years of trial and error whose well-being now depends on us.

The good news is that we are increasingly seeing the world in this way. Science has awakened us to the remarkable intricacies, permutations, and pure-luck events that came together to bring forth life on this third planet from the sun. The abundance and variety of life that has populated this world over time is staggering. Many cultures before us surely marveled at the world they inhabited. All the more so do we, for we can see what they never could. Our archaeology allows us to peel back the layers of millions of years, offering both hope and cautionary tales about the future.

Our telescopes allow us to peer across the unfathomable vastness of space to the beginnings of time, into a universe thirteen billion years old, with two trillion galaxies, two hundred billion trillion stars, and more planets than we can count. And yet in all of that, we have failed to find evidence of life anywhere else. As far as we know, we are it. Even if there is life somewhere out there, or if there had been life sometime out there, or if in ten billion years there might again be life somewhere, we might never ever know. So we are effectively alone. For me at least, that awestruck, gobsmacked sense of our planet is one of the best motivators for cherishing and saving it.

Not everyone alive today is equally culpable for the

earth's degradation. Many of us who are baby boomers, raised in America in the most gilded and profligate era of humankind, have been the greatest beneficiaries of the fossil fuel era. Boomers (like me) have lived in the heyday when fuel was so cheap that gas stations gave away plastic dinosaur toys and sets of dinner glasses to attract and reward their customers.[13] The highway system that connects the states, agriculture's Green Revolution, which utilizes chemicals to enable the worldwide explosion of food production, the use-once-and-throw-away economy—all grew up with the boomers. We were raised believing the world is overflowing with abundance, ripe for the picking, full of endless possibilities. We never even thought to worry about the stability of the climate or the security of the seasons or to wonder if we could trust in the earth's capacity to provide us with what we needed and receive what we tossed back to it.

The biggest discrepancy in contributors to environmental degradation is not between the generations though. It is between the wealthy and the poor. According to the United Nations, "the combined emissions of the richest one percent of the global population are larger than the combined emissions of the poorest 50 percent."[14] The many are suffering from the profligacy of the few.

But we know now that the practices of the last century that seemed to enhance our lives are robbing our children and grandchildren of theirs. Those of us who thrived most from this self-induced environmental crisis must be the most committed to righting it. Those who are younger, who are inheriting this compromised world through no fault of their own, look to the future with fear and to us with a sense of betrayal. They are urging us to work, full throttle, to make things right.[15]

We can do this, together. It is not (yet) too late. With our collective will, across the generations, we can muster the attitude that will enable us to build the systems that can support us all. Thomas Berry said that every generation has its "Great Work," a task that they did not ask for, a task they might never have chosen, but one they are nonetheless called to do.[16] The Great Work of today's generations—all of us alive today—is nothing short of saving the earth. And to do that, we need a revolution in values, a renewal of spirit that can respond to the urgency of the moment.

In an odd way, we have been handed a most humbling privilege: to be alive at this pivotal moment and entrusted with determining the fate of earth and humanity. Our goal of establishing a thriving world for all Creation must now become not just our task but our devotion.

The People of the Land: Jews and the Earth

Jewish tradition is built upon just such devotion. It is born from our relationship with the land—both the Land of Israel, which is the land of the Jewish people, and the land of all the earth. The former, fraught with history, memories, intimacies, laws, and sacred encounters, has schooled us about the latter. As much as we are the people of the book, *am hasefer*, we are also the people of the land, *am ha'aretz*. Our founding texts, practices, holidays, and calendar are all grounded in the story of our relationship with land. Connecting through land is the origin story of the Jewish people. We were called into peoplehood through a call to land: "God said to Abram, 'Go forth from your native land and from your father's house to the land that I will show you'" (Genesis 12:1). God blessed the Jewish people with the fertility of the land, and the Jewish people thanked God by returning gifts of the land. "When you enter the land that your God is giving you as a heritage, and you possess it and settle in it, you shall take some of every first fruit of the soil, which you harvest from the land that your God is giving you, put it in a basket and go to the place where your God will choose to establish the divine name" (Deuteronomy 26:1–2).

The whole of Torah—the Five Books of Moses—is the saga of the Jewish people heading to their land and the promised fertility and goodness of the land. It is through this story, which speaks of our particular relationship to a particular parcel of land that the Torah spins out an ethic that teaches us how we should behave to all the earth.

Land in the Bible was always more than soil and space, even more than the particular homeland for the Jewish people. It was an active presence, a medium of God, a character with agency in our sacred story whose needs we were called to recognize and honor. Land served as a bridge between the human and the Divine, a way for God to communicate with the Jewish people and for the Jewish people to communicate with God. Before the onset of synagogues and communal prayers, before the study of sacred texts became an expression of piety, God and the Jewish people spoke with one another through the language of nature. Rain and soil, animals and crops, tithing and sacrifices, in short the bounties of the earth shared between heaven and earth, became the language of love between God and the Israelites.[17] The Israelites' sense of time, seasons, morality, philanthropy, piety, and identity were all expressed and measured through the ways they engaged with the land.

So, in a way, it is for all of us today, for the Bible is both an artifact of its times and a book for the ages. To be relevant over the millennia, it had to be read creatively and interpreted deeply, in concert with the needs of the times. Today, biblical lessons of caring for the land of Israel can serve as an ethical manifesto governing our behavior for all land everywhere. Read that way, and through the lens of two thousand years of folk and rabbinic interpretation, a Jewish land ethic begins

Over the centuries, the phrase *am ha'aretz*, "people of the land," took on a derogatory cast. It was used as a term of disparagement, denoting the boorish, uneducated person at worst, the simple commoner at best. Such was the impact of years of exile and movement and Judaism's alienation from the land. But as the First and Second Aliyah movements sought to reclaim farming in Israel over the late nineteenth and early twentieth centuries, as the kibbutz movement tied its members to the land, as the Jewish people worldwide are reclaiming their relationship with the earth, it is time we reclaim this phrase, bolstering our identity as the People of the Book.

A land ethic is a call to care for a place inspired by knowledge, intimacy, aesthetics, appreciation, and humility for all that the land does and offers. The naturalist Aldo Leopold famously crafted his now-classic land ethic this way: "A thing is right when it tends to preserve the integrity, stability and beauty of the biotic community. It is wrong when it tends otherwise."[18] This book is an effort to uncover the ideas and vocabulary of a Jewish land ethic that can guide the ways we live.

to emerge—an ethic that is defined by mutual care, an ethic that is a covenant of companionship between people and the land.

Just So You Know: Defining Key Terms and Our Approach

Before we jump in to explore this covenant, we need to clarify two things.

First: language. I use the word "land" in two ways. Sometimes I use it in its more literal meaning of that specific domain that we trod on: soil, fields, rocks, etc. Other times I use it to mean all the above plus water, air, animals, plants, insects—in short, the entire biotic community. "Land," though, is a much more compact way to say all that.

Sometimes, too, I will use "nature" where I could have used "land" in its expansive meaning. But nature will tend to be broader and include operating systems that transcend the biotic community. So, too, with "earth." Sometimes it will just refer to the material elements and other times include the systems that guide them. The word "life" will sometimes refer to all that, the vibrancy of the material world around us, the plants, animals, and systems that sustain them, and sometimes it may mean the very idea of Creation, being, existence itself. Since these meanings often flow into one another, usage will be a bit fluid. But hopefully the context will make things clear.

Second: an explanation of how and why I bother to read the classic Jewish texts. The Hebrew Bible (the *Tanach*), the Talmud, and midrash are Judaism's foundational texts. They are the sacred literary legacy of the Jewish people and ground our spiritual seeking. But they do so in different ways for different people.

For those whom we might call traditional believers, the Torah, the text that began it all, has the immutable authority of God. Everything in it—every letter, every pen stroke—harbors an array of sacred meanings. Its lessons are boundless. As Ben Bag Bag, a rabbinic sage of the first century CE, said, "Turn it, turn it, for everything is in it" (*Pirkei Avot* 5:22).

The Talmud, on the other hand, everyone agrees, was crafted by humans. It is the effort of the rabbis over a five-hundred-year period to unpack and apply the eternal message of Torah to their contemporary times. That is an effort that never ends. So it is the sacred task of each generation thereafter to interpret those original words in line with the ideas and laws appropriate for its days.

I, for one, am not part of the faithful who believe in the divine authorship of the Torah. Nor do I adhere to the intractable authority of the early rabbis. Yet I am deeply devoted to these texts and their teachings. I believe they are imbued with an air of the sacred. They preserve the most sublime struggles of my people as they sought to make sense of this conundrum of a world, distilled through the experience of the ages. The good, the bad, and the ugly in this struggle are mine. I claim my ancestors' experiences as my inheritance. They serve as guides that help me navigate the wilds of this one shared, improbable human adventure.

Throughout this book, when recalling traditional texts and beliefs, I will often speak in the words of tradition and present the sacred story as its authors tell it. Know that underlying it all, I understand the text as a kind of metaphor, a way of giving authority to values that merit our consideration and possible allegiance. I do not believe in the truth of all that tradition tells us.

Tanach is an acronym made of the first letters of the names of the three major sections of the Hebrew Bible: **T**orah (Genesis, Exodus, Leviticus, Numbers, and Deuteronomy); ***N****evi'im*, the Prophets (such as Isaiah, Jeremiah, Ezekiel, and the minor prophets); and ***K****etuvim*, the Writings (such as Psalms, Proverbs Esther, and the Song of Songs). The Talmud is the majestic legal compendia composed in the rabbinic academies of Israel and Babylonia over the course of six hundred years (roughly 0–600 CE) that record the deliberations, disagreements, and delights of the earliest generations of rabbinic thought. The midrash, unlike the Talmud, is a genre more than a book. It comprises a sweeping array of rabbinic commentary and imagination grounded always in the *Tanach*.

Know too that when I use the word "God," I do not mean "God" the way our ancestors did. For me, references to "God" speak to the collective dreams, fears, and wisdom that our ancestors wove into a divine personality. I use "God" and mean the dream-song of our tradition.

With all that, the texts still speak powerfully to me. For they are my ancestors' earnest, soulful efforts to find answers to the hard questions of life. In them, I find reflections of hundreds of generations of Jewish hopes, fears, and efforts to get this Life Thing right. For their days—as ours—were framed by the frailty of the human spirit, the opaqueness of life's purpose, and the temptations of unfettered desire.

Knowing that, their teachings amaze. How many of us would choose to help our enemies as the Torah bids us? ("If you see the donkey of someone who hates you fallen down under its load, do not leave it there; be sure you help them with it" [Exodus 23:5].) Who among us would regularly let the poor onto our land to take our hard-won harvest? ("When you reap the harvest of your land, you shall not reap all the way to the edges of your field.... You shall leave them for the poor and the stranger" [Leviticus 19:9–10].) And who of us truly, wholeheartedly loves our neighbors as ourselves (Leviticus 19:18)?

To know that the authors of our sacred texts could have succumbed to the same selfish desires that tempt most of us but nonetheless rose above them is enough to earn my respect and attention. To know that for thousands of years, one generation after another preserved and elaborated on the texts of their ancestors, though they could have walked away, captures my curiosity. What is in there that has claimed such allegiance?

There is nothing particularly surprising when a tradition that believes in divine authorship tells us that the word of God calls us to be good. But when the word was written by mere humans, it is different. We have faults, we have frailties, we have temptations. And when, despite our ancestors' existential struggles with futility, hedonism, and selfishness; despite their personal and collective battles with fear, vulnerability, and the seduction of nihilism, they nonetheless chose to craft and preserve a tradition that is dedicated to the pursuit of goodness, they have earned my admiration. I will study their teachings eager to be inspired by them and make them my own.

Even Torah's difficult texts offer insights and guidance. For if we dig beneath the surface and excavate below their words, we discover the universal questions that we all struggle with. We should read their texts as answers to those questions, as their attempts—if sometimes faulty and far from laudable—to comfort themselves and contain the fear of disorder as they face life's existential struggles. And in reading the texts this way, we hold a mirror up to ourselves and can hardly avoid asking, how are we doing?

And so, we begin.

CHAPTER TWO

 The Very First Command

"When God began to create heaven and earth..." (Genesis 1:1). These first words of the Torah are so familiar that we tend to take them for granted. But we shouldn't.

The Torah could have opened in any number of ways: with exultations and adorations of God, celebrating God's holy and wholly-other heavenly sphere, or with God's vanquishing of other gods and powerful forces of nature (there are hints of such stories scattered throughout the psalms). It could have opened with the story of Abraham and the particularity of the Jewish people or the subtle presence of the burning bush. It could have started with the explosive Revelation of the Torah at Mount Sinai or the conquest of the Land of Israel—or something else entirely.

Yet instead of beginning in any of these ways, the Torah chose to begin with God fashioning the fullness of Creation, encompassing all then-known matter and manner of life. It chose to begin large, with a story that embraces not just the Jewish people but all peoples, along with all creatures, all existence. The reader enters the text and immediately is transported into the advent of time and space, being and blessings, and the unfolding mystery of life here on earth. The six days of Creation are laid forth before us, showing us God speaking the world into being. "And God said: Let there be light.... And God said: Let the waters below be gathered together.... And God said: Let the earth sprout forth seed-bearing plants" (Genesis 1:3, 1:9, 1:11). We see the world take shape, step by step. Even more, when we read closely, we see beyond the what and how of Creation and catch a glimpse into the why. We are here, the text seems to simply suggest, because that is what God wants.

Chapter 1 of Genesis tells the story of the systematic and ordered creation of the world. Day by day, God fashions the world, each day serving as the foundation of the next.
Day one: the audaciousness of light
Day two: the separation of heaven and earth
Day three: the gathering of the earthly waters to reveal dry land, allowing for the emergence of grasses and trees
Day four: the appearance of the orbs of the heavens
Day five: the creatures of the seas, rivers, lakes, and skies
Day six: the land animals and the humans
Day seven: rest.

That is: God desires the ferocious chutzpah of being.

God does not charge humanity here with any work (that comes in the next chapter). God does not give us trials to test our worthiness. Rather God gives us blessings and visions of abundance and sends us forth to enjoy.

This is important because origin stories are more than stories of how things began. They are stories of orientation, of purpose, of teleology—stories that hint at who and how we are supposed to be. This origin story provides us with a vision of a world full of goodness and vibrancy, a world that celebrates the presence of life.

Creation is seen as the expression of God's desire, the intentional, thoughtful, deliberate act of God choosing life. For what divine purpose? We don't know. But the story tells us that Creation was no accident, no afterthought, no serendipitous celestial whittling designed to fill eternity's boredom. The world in both idea and execution was desired and pleasing to God. After each step of Creation we read that God surveyed the divine handiwork to see how it looked and functioned: "And God saw." And after each surveyance God rendered a decision: it is good.[19]

More than an assessment of technical achievement, this was a pronouncement of sacred worth. Creation, Torah is telling us, is a thing of beauty, integrity, generativity, value. Each and every stage of Creation is good. And when the discrete parts came together at the end of the sixth day to form one united whole, when the entire system clicked and hummed and in sum became more together than they were apart, it was more than good. "God saw all that had been made and found it very good" (Genesis 1:31).[20]

In short, the Torah is urging us to see, first and foremost, that life is awesome.

Life exists for the simple, glorious purpose of existing. God chose existence over emptiness, order over chaos. Life's goodness and meaning are baked into its existence. The astonishing array of atoms and stars, crickets and dinosaurs, joy and pain, yesterday and tomorrow, the rain, the rainbow, and the pot of gold at the end are all rolled into one celebrated explosion of being. The goodness inherent in Creation is, in a word, self-evident. Its purpose is none other than the glorious, improbable fact of being and the continuation of being until the end of time. As the modern Jewish philosopher Lenn Goodman writes, Creation "is a good thing, good in itself, a thing that should exist, that deserves to exist, for the beauties it now bears, through no prior claim, but for its value now… in itself… intrinsically."[22]

One needn't be a believer to be swept up in this vision. Life is miraculous and precious, no matter its origin. And once it begins, it urgently, constantly, seeks more life.

We humans are part of this grand parade of being, a wondrous aspect of this ever-changing, yet ever-enduring world. We are bound to the waters and bound to the air, bound to the plants and lichens and invertebrates, bound to bacteria, fish, and stones. But, as much as we are part of Creation, Jewish tradition tells us we stand apart from it too. For we possess great knowledge and can use it to exercise self-determination—which in turn endows us with great responsibility. Maimonides, the preeminent twelfth-century Jewish philosopher, writes, "We are unique in the world…able to discern between good and evil, we can do whatever we desire. Nor is there anyone to prevent us from doing good and

Rabbi Ellen Bernstein, a pioneer in the modern Jewish environmental movement, wrote, "The idea that a God exists who created heaven and earth is truly profound. It means that the earth that we walk upon, the air that we breathe, the food that we eat, are all signs that the world is filled with mystery. Those who cherish this idea sense that everything they encounter is sacred. Nurture this idea, and it will guide the choices you make and the way you live your life."[21]

Life can be found in the most unlikely of places. One gram of sticky black goo in aptly named Pitch Lake in the Caribbean can harbor up to ten million microbes. Amid the extreme heat and pressure of hydrothermal vents in the depths of the ocean—a place once thought to be devoid of life—tubeworms, clams, and shrimp abound. Microbes are found in the Dead Sea (disproving its name). And of course there are the irresistibly adorable, seemingly indestructible tardigrades, affectionately known as water bears, which can withstand searing heat, extreme cold, and even high amounts of radiation and remain unharmed.[23]

evil. . . . This means the choice is in our hands."[24] Humans, distinct from the rest of Creation, know and can turn that knowledge into action, for good or bad. We can assess alternatives, decide what to do, choose how to behave, which means we are uniquely responsible for our behavior and accountable for its consequences.

Today we can affect the world as never before and in ways we cannot always predict. What we choose to do can make the difference between life and death. So we must choose carefully. Tradition urges us to choose the preference for being, for life. Speaking in the name of God before all the universe, the Torah says, "I call the heavens and the earth as witnesses against you that I have set before you life and death. . . . Therefore, choose life" (Deuteronomy 30:19).

The imperative to choose life presses on us for another reason as well. We are the ones, perhaps the only ones, who know that the universe exists. Therefore we are the ones—perhaps the only ones, anywhere—who are sentient enough to acknowledge and celebrate it. As Thomas Berry teaches, we are the consciousness of the universe, the ones who are witness to the magnificence of it all. How sad, perhaps pointless, it would be if the universe's pulsing, glowing, bursting, and twirling thrummed on for billions of years without anyone anywhere ever noticing; without anyone telling stories of its beginnings, drawing pictures in the stars, unlocking the secrets of the galaxies; without anyone looking up and cheering the pure audacity of Creation.

Our job, then, as both beneficiaries of and witnesses to the miracle of Creation, is twofold. We are to cherish and rejoice in this exuberance of life. And we are to care for it and be as beneficent to it as it has been to us.

A Commandment Called Yishuv Ha'olam, *Preserving a Habitable World*

Before the beginning there was lifelessness, a world bereft of the vibrancy of being. "When God began to create the heaven and the earth, the earth was unformed and void [*tohu vavohu*]" (Genesis 1:1–2). It is hard to translate this dark lyrical phrase. "Wild and waste," is how Everett Fox translates it. "Welter and waste," per Robert Alter. "Swirling chaos," "meaningless or vacuousness," "pure entropy" could all work. The words conjure up a vision of the universe before Creation, the epitome of disorder, nothingness, or worse, meaninglessness. Only when God imposed order, creating distinctions in matter, establishing light here, sequestering darkness there, did the conditions exist for life to emerge. Today we might say, only when protons and neutrons began organizing themselves into the earliest atoms, only when the gaseous masses after the big bang cooled sufficiently to allow the stars to form into distinct clusters—only then did Creation as we know it begin.

Order is an essential ingredient of life. Order provides the necessary, reliable, stable environment that brings forth and sustains life. Plants and animals depend on the reliability of the seasons, the course of hot and cold, the cycles of sun, the patterns of the rain. The tides, the winds, the ocean currents all mark the life-giving rhythm of time. We humans may be godly, but we cannot conjure up or decree nature's order. Only powers beyond our ken can do that. But we are godly enough to affect this order. Our actions can encourage or disrupt it. Our very first task, then, the first call to humanity, is to protect this order and the elements of the world that support it.

The phrase "choose life" must not be confused with "pro-life." Choosing life in this context means to embrace a life of morals and good deeds. In the language of tradition, it means to embrace a life of Torah, so that all we touch is blessed by our being. Choosing life means that we recognize the power of our actions, both personal and collective, and responsibly exercise the wisdom that this power demands of us.

Thomas Berry was an insatiably curious Catholic priest, a student of world religions, and one of the most influential religious naturalist thinkers of our time. Calling himself a "geologian," he urged humanity to embrace what he called our Great Work: the need to redesign our relationship with the physical world. "Our challenge," he wrote, "is to create a new language, even a new sense of what it is to be human."[25]

Order is not the same as stasis. Life needs its edginess, its diversity, its ability to explore, evolve, and renew. Experimenting with change, mixing things up a little, allows life to survive when systems change or undergo cataclysmic events. Life needs pioneers.

It should come as no surprise, therefore, that it is just this task, this essential mandate, that is captured in Judaism's very first mitzvah (one of the 613 commandments based on the Torah): *yishuv ha'olam*, the call to establish, preserve, and maintain an ordered, habitable world here on earth. It is upon this mitzvah, and its successful pursuit, that all we know, all we are, all we love, depends.

This mitzvah is the precursor of everything else to come, a far-reaching positive command that is designed to guide all our acts. It is grounded in the narrative of renewal in Genesis 1, that glorious rehearsal of Creation's regenerativity. And it is picked up by Isaiah as a foundational teaching.

"God," Isaiah writes, "is the Creator of the heavens, the One who fashions the earth and makes it. God did not create the earth to be chaos [*tohu*]. God fashioned it to be full of life, to be inhabited [*lashevet*]" (Isaiah 45:18). Isaiah exults in the creation of the world and God's intentions for it. The world, he asserts, was made to be purposeful and enduring, not accidental and transient. It was to be full of life and not *tohu*. *Tohu*, the same word found in the beginning of Genesis 1, cannot bring forth life, it cannot bring forth beauty, it cannot bring forth meaning. *Tohu*, we saw, is emptiness, waste, and welter. We dare not return there, though the specter of *tohu* hovers over us. Life is its defiance, its antidote. The Jewish tradition teaches that God created a world of and for *yishuv* (habitability) in radical opposition to *tohu* (nothingness). And God charged humanity with preserving the order and habitability that would sustain this life-studded world.

Rashi, Rabbi Shlomo Yitzchaki, the premier eleventh -century sage and commentator of both the Hebrew

Bible and the Talmud, amplifies this call to preserve life. He explains that this verse from Isaiah is not just *descriptive* of God's intent but *prescriptive* of our task. "We should continually occupy ourselves with the establishment of the world," he writes. God made the world and handed its care over to us. Consonant with the enormous reach of our capacities comes the obligation to be constantly engaged in "the preservation of the world."[26] Jewish tradition's claim that humans are made in the image of God confers on us both great powers and great responsibilities. As one commentator explained, to believe that we are made in the image of God defines our life's work, which is to do good—in the service of all.[27]

That is what the mitzvah of *yishuv ha'olam* asks of us: to do good for the purpose of all life. There are different kinds of mitzvot. Some are prompted by the clock or calendar—like blessing the new moon or lighting Shabbat candles. Others require waiting for the opportunity to present themselves—like entering a boy into the covenant on his eighth day of life or blessing the bread before eating. *Yishuv ha'olam* is a different sort of mitzvah. There is never a time it cannot be pursued. It is implicated in all our acts and decisions, for almost everything we do affects the natural world and thus life all around us. The crops we grow; the foods we eat; the energy we use; the ways we build; the stuff we buy; the things we discard and the ways we discard them all affect the health of the land and the well-being of the world. As a result, "There is no greater act of righteousness than this," writes Rabbi Mordecai ben Avraham Yoffe, the Levush, who lived in the sixteenth century: "to repair and maintain the habitability of the world [*tikkun v'yishuv ha'olam*]."[28]

The incomparable Carl Sagan poetically and powerfully expresses our responsibility to our planet. In his 1994 book *Pale Book Dot,* he wrote about a photo of earth taken by the Voyager I space probe from 3.7 billion miles away: "Look again at that dot. That's here. That's home. That's us. On it everyone you love, everyone you know, everyone you ever heard of, every human being who ever was, lived out their lives. The aggregate of our joy and suffering, thousands of confident religions, ideologies, and economic doctrines, every hunter and forager, every hero and coward, every creator and destroyer of civilization, every king and peasant, every young couple in love, every mother and father, hopeful child, inventor and explorer, every teacher of morals, every corrupt politician, every 'superstar,' every 'supreme leader,' every saint and sinner in the history of our species lived there—on a mote of dust suspended in a sunbeam."

The root of the word *lashevet* yields the word *yishuv*, meaning "the place or act of establishing a settlement, a home, a collective vibrancy." *Olam* is the word for "world," the all-encompassing entity in which we exist. Hence, *yishuv ha'olam* means "the preservation of the habitability of the world." While some classic rabbinic texts seem to restrict the application of *yishuv ha'olam* to the realm of gainful employment or contributing to a productive economy, the exigencies of today demand that we expand the meaning of the words to embrace the full imperative of caring for the habitability of the world in all ways.

Over a thousand years ago, an anonymous teacher captured the primacy of *yishuv ha'olam* in this cautionary text: "Just after creating *adam*, the first human, the Holy One took him around to all the trees of the Garden of Eden and said to him: See how beautiful and wonderful My works are. Everything I have created, I have created for you. Be mindful that you do not ruin and devastate My world, for if you ruin it, there will be no one after you to set it right."[29]

We are permitted—even encouraged—to use the resources of this world in ways that enhance our dignity and enjoyment as long as it also enhances the earth's enduring habitability. We can build homes from its wood, fashion towers from its metals, hew blocks from its granite. We can harvest flax to make linen, spin wool into yarn, use our ingenuity to make materials nature never dared. We can farm fields so they increase the fullness of their yield. But amid all this creativity, we must remember our charge. We now realize we must do all this in ways that increase the world's generativity and do not diminish or destroy it.

Our Urgent Spiritual Call: A Habitable World

It seems curious, then, given the essential nature of the mitzvah of *yishuv ha'olam*, that it has been so little known to so many Jews. Perhaps it was overlooked because it is not found in the categorical list of 613 commandments compiled by Maimonides.

Perhaps, too, it was overlooked because most previous generations could not have easily grasped that humanity would ever become a "geophysical force," as the famed Harvard ecologist E. O. Wilson describes humanity today. For mitzvot to be activated, they must respond to the needs of the age. It was hard to imagine before the twentieth century that humanity could irreversibly upend the globe's operating systems. The ancients knew that we could despoil rivers and deplete pastures and fields, create localized famine, and force the dislocation of whole populations. But changing the seasons, melting the ice caps, causing entire species to become extinct? That was an understanding that was born and came of age in modernity.

Two hundred years after Maimonides compiled his list of commandments, the author of *Sefer Hachinuch* (The Book of Learning) set out to craft a somewhat different list, both in order and content. While Maimonides, always the organizer and logician, arranged the mitzvot by topic, the author of *Sefer Hachinuch* arranged them by the order in which they appear in the Torah. He counted the mitzvah to "be fruitful and multiply," found in Genesis 1:28, as the very first, not just in the narrative order, but also in theological and substantive importance.

To *Sefer Hachinuch*, the words "be fruitful and multiply" mean more than they first suggest. He understands them to refer not to procreation per se but to the grand enterprise of the habitability of the world, *yishuv ha'olam*. Though procreation is essential for the continuity of life, he explains, procreation alone, without caring for the well-being of the world, does not fulfill the command. To bring children into the world denuded by our acts, contaminated with pollutants, and beset by climate change does not fulfill the command to be fruitful and multiply. *Yishuv ha'olam*, the healthy habitability of the world, not procreation alone, is "the foundational mitzvah on account of which all other mitzvot in the world exist."[30]

Sefer Hachinuch was not alone in saying this. He was expanding on a teaching that was articulated three

Just as *tikkun olam* (repairing a broken world) is taken to be a fundamental tenet of modern Judaism and is known both within and beyond the Jewish community, so may it be with *yishuv ha'olam*. Over the next decade, in response to the imperative of caring for the earth, may this phrase flow easily and often from our lips, guiding our actions and (re)claiming its place as a fundamental tenet of our tradition.

In classic Jewish tradition, authors are often called by the title of their best-known book. So we have the Chofetz Chayim, after his book on the ethical use of language, whose name is Yisrael Meir Hakohen Kagan. Or the Levush, after the book *Levush Malchus*, a ten-volume set on the laws and customs of East European Jewry, whose name was Mordecai ben Avraham Yoffe. So too the author of *Sefer Hachinuch* is called by that name, especially since the authorship of the book is contested.

hundred years earlier. "Some are of the opinion," wrote Saadia Gaon, the prominent tenth-century Babylonian rabbi and the premier rabbinic authority of that time, "that people ought to dedicate themselves to begetting of children.... I say, however, of what benefit are children to a person if he is unable to provide for their sustenance, clothing, or shelter? And what is the good of raising them if it will not be productive of wisdom and knowledge on their part? And of what use are the pity and sympathy lavished upon them in the absence of these factors unless it be to add to the heartache of the parents?"[31] While Saadia was likely speaking about one's individual responsibility of parenthood, today we can expand this teaching and apply it to our collective responsibility. For through our actions, we all contribute to the conditions in which the world's children will grow up.

Procreation, therefore, is not a goal in itself. It is a necessary but insufficient response to the call for *yishuv ha'olam.* All children deserve the opportunity to pursue their dreams in a world at peace with itself, with seasons that are predictable, a climate that is habitable, lands that are fertile, waters that are teeming with life. All of these are dependent on a world capable of stabilizing and renewing itself. These factors make for a dignified life. These are the demands of *yishuv ha'olam.*[32]

Everything we experience, *Sefer Hachinuch* seems to be reminding us, everything we do, everything we cherish, from apples to zippers, depends upon a world that functions well. And a world that functions well now depends on us. Absent our concerted efforts to right the wrongs we have wrought, the forces of chaos we have unleashed will continually seek to pull things apart, to diminish and wreak havoc on the world around us. Our task is to hold things together, to join our genius with the workings of the world to provide for a stable, fertile, trustworthy ground of life. "The aim of the Torah is this: that existence should continue existing," so wrote Isaac ben Judah Abarbanel in the fifteenth century.[33] It sounds so simple, so obvious. Until it isn't. A tautology that hardly needs to be said. Until it does. The mitzvah of *yishuv ha'olam* is a critical response to the urgent call to preserve this habitable world.

Hans Jonas was a prominent twentieth-century Jewish philosopher who devoted much of his work to warning that technology, with all its blessings, could easily threaten the well-being of the world—and us. We must be more humble, he argued, in deploying technology than we have been until now, for we are often unaware of its unintended consequences. We dare not play dice with the well-being of the earth, introducing changes that can cause irreparable damage. Our progress should advance the way nature advances, he urged, making small changes, here and there, testing new models without destroying the whole. We should gain no consent to deeds that will harm the humanity of the future. Rather, we have an "unconditional duty" to ensure the continuity of humanity. "Never," he wrote, "must the existence or the essence of [humankind] as a whole be made a stake in the hazards of action."[34] We must proceed with creativity restrained by caution. We must act, he calls to us in the voice of a prophet, so that the effects of our actions "are compatible with the permanence of genuine human life."[35]

While Jonas crafted his argument centered on humans (not unusual for his time), we can expand his thinking to encompass all Creation. And imagine that he too, were he alive today, would agree. For there is no refuge for humans on a despoiled, empty earth. To say that we should seek to make all our actions be compatible with the permanence of life is as fine an encapsulation of the mitzvah of *yishuv ha'olam* as one can find.

The Value of a Commandment

The idea of touting a commandment may seem odd—even counterintuitive—in a society devoted to personal liberty. Yet for many, commandments in the Jewish imagination are covenantal, not coercive. Following them is a choice, an expression of love and commitment, not blind obedience, just as responding to the request of a loved one is an act of love and commitment.

And commandments elevate. For embedded in the idea of a covenantal commandment are three core beliefs: the celebration of the inherent worth and dignity of each individual, the recognition that we are bound to one another in covenantal relationship, and the understanding that what each of us does in this world matters. Commandments remind us that we are not insignificant, not overlooked, not alone. Rather we are part of a whole, a partner in a larger story, and our actions resonate far beyond our personal sphere. To accept the concept of a commandment—even as we might reject the concept of a Commander—places us in the midst of a community of belonging, of meaning. It offers us a guiding ethic in which who we are and what we do are woven into a larger whole. Accepting the idea of commandment is to live in a culture of "we," not just "me."

Such consciousness can serve as a welcome antidote to the feelings of smallness, loneliness, and futility that might wash over us in today's world, especially its overwhelming environmental challenges. Many of us fear that individually our singular actions are insignificant, that our gestures at sustainability are theater at best and futile at worst. Feeling that way, we may ask: Why bother? Why care about reusable shopping bags? Why limit how much meat I eat? So what if I don't recycle? Why go to the trouble of divesting from big banks that support fossil fuel extraction? What good will one more protest do?

The frame of "commandments" teaches that our actions do make a difference—both for the world and for us. *For the world* because all good causes start with a small band of devotees. The famed anthropologist Margaret Mead is remembered as saying, "Never underestimate the power of a small group of committed people to change the world. In fact, it is the only thing that ever has."[36]

Judaism teaches us that we are to pursue the good cause, no matter how small our ranks. It is not up to us to win. It is up to us to try. Two thousand years ago, Rabbi Tarfon said, "It is not your obligation to complete the work, but neither can you neglect it" (*Pirkei Avot* 2:16). We must get in the game. Nonaction will not get us a win. Small acts may inspire ever more acts, which may cumulatively grow and attract even more. When our good efforts are joined with others similarly motivated, a movement will inevitably flourish.[37]

For us because we become the things we do. Meaning and purpose emerge from doing. Our beliefs and sense of self follow from our actions even as our actions are inspired by our beliefs. Modern psychology teaches us that we are uncomfortable when our actions and beliefs are at odds with one another and that we often resolve this discomfort by adjusting our beliefs to align with our actions. Jewish tradition teaches likewise: "You must know that people respond according to their actions. Their hearts and all their thoughts always follow after the actions that they do—whether good or bad... for hearts are drawn after the actions."[38]

We become what we do. Knowing this, Judaism seeks to guide behavior. It celebrates the incremental, the power that small acts of goodness have to gradually make of us the person we want to be. Tradition teaches that the observance of one mitzvah leads to another, then another and another (*mitzvah goreret mitzvah*), creating a better me, who influences a better you, and through us, makes a better world.

A New Category of Commandments: Between Us and the World

To guide us in our choices, Judaism organizes the commandments in two general categories: ritual commandments, which it calls "commandments between humans and God" (*mitzvot bein adam laMakom*), and ethical commandments, which it calls "commandments between one person and another," (*mitzvot bein adam l'chaveiro*). The former include prayer, Shabbat candle-lighting, laws of kashrut, etc., while the latter include redeeming the captive, loving your neighbor, being honest in business practices, and more.

Some modern Jewish environmental advocates, however, premier among them Jeremy Benstein, are suggesting that we need a new, third category of commandments guiding the actions between us and the earth. These are alternately called *mitzvot bein adam l'olam* (commandments between us and the world) or *mitzvot bein adam l'adamah* (commandments between us and the land).[39] Such a category would include actions both new (such as using renewable sources of energy, implementing clean transit systems, eating a more plant-based diet, practicing regenerative agriculture, reimagining plastic, composting waste from everything that grows) and old (such as not wasting, creating green spaces throughout and around cities, protecting the commons). Organizing the commandments this way would clearly, explicitly integrate caring for the earth into the sacred Jewish enterprise and would be a practical expression and application of *yishuv ha'olam*.

The Time Is Now

Our task is clear. Modernity is calling us to reimagine our relationship to nature. We know what science—and the seasons, the oceans, the floods, the fires, our household thermometers—are telling us. We know how to retool our economy so that we thrive in systems that can seamlessly support a healthy earth and a fulfilled humanity. We know we have to align public policy, the marketplace, our own hearts and appetites to be in sync with the rhythms of the natural world.

Each generation is charged with handing a healthy world on to the next (especially the boomer generation that benefited so greatly, and now we know recklessly, from the fossil fuel era). No one generation may lay greater claim to the treasures of the earth than any other. No one nation or continent or corporation or species can appropriate earth's resources for its own benefit to the detriment of others. Across time, across space, we are in this together. We must all work to right the wrongs we have caused. The only thing is, we dare not despair.

Rabbi Daniel Swartz, a longtime advocate of Jewish environmental engagement, writes, "To succumb to inaction because the problems we face are complex,

The commons is an idea steeped in history that is having a resurgence. It is "those things that we inherit and create jointly, and that will (hopefully) last for generations to come. The commons consists of gifts of nature such as air, oceans and wildlife as well as shared social creations such as libraries, public spaces, scientific research and creative works," explains On the Commons, an organization devoted to reclaiming the appreciation and practice of the commons. It is those aspects of life—both natural and cultural—upon which we all depend and no one in particular can claim as theirs alone. And it is a concept that is still very much alive. In 2009, Elinor Ostrom won the Nobel Prize in Economic Sciences for her work on the way of the commons.[40]

Retooling our economy is within our grasp.

Retooling our economy is within our grasp. Modern economists such as Herman Daly, Kate Raworth, and others have been urging us for decades to pursue models of a sustainable economy. Including "externalities" into the accounting of an object's cost, requiring producers to be responsible for disposing of or recycling their products and packaging, taking into account society's well-being as a component of GDP, quantifying the value of ecosystem services such as pollinators, trees, carbon sequestration—all can help us redesign how we structure our economy. The Wellbeing Economies Government partnership is a group of countries pooling their wisdom and experience to craft such economic models that account for social well-being. As defined by the Wellbeing Economy Alliance: "A Wellbeing Economy is an economy designed to serve people and planet, not the other way around."[41]

Robert F. Kennedy, brother of President John F. Kennedy and champion of human rights, spoke of the limitations of our accounting system in a speech he gave on March 18, 1968. In it he said:

> Too much and for too long, we seemed to have surrendered personal excellence and community values in the mere accumulation of material things. Our Gross National Product . . . counts air pollution and cigarette advertising, and ambulances to clear our highways of carnage. It counts special locks for our doors and the jails for the people who break them. It counts the destruction of the redwood and the loss of our natural wonder in chaotic sprawl. . . . Yet the gross national product does not allow for the health of our children, the quality of their education or the joy of their play. It does not include the beauty of our poetry or the strength of our marriages, the intelligence of our public debate or the integrity of our public officials. . . . It measures everything in short, except that which makes life worthwhile.[42]

because our ideals are challenging, because there is pain along the way, is to abrogate our partnership with God in creating a better world, to abandon our stewardship along with our ideals, along, finally, with our humanity."[43]

The task is daunting, for the stakes are high. But the earth, the future, and those yet to come are depending on us. From the beginnings of Genesis to the twenty-first century, caring for the land has been part of Jewish tradition. The mitzvah of *yishuv ha'olam* is emerging, ascending, assuming its rightful inspirational place to guide us in the task ahead.

CHAPTER THREE

Survivability Versus Sustainability

To want to make this world a habitable place is one thing. To know precisely how to do it is another. For that we need a vision, a direction, a blueprint, a user's manual.

Fortunately, the Torah offers us just that: two versions, in fact, two stories of Creation—one in Genesis 1:1–2:3 and a second in Genesis 2:4–3:24—that lay out such guidance.[44] These two Creation stories present two different visions of Creation, two different views of nature, and hence two different roles for humanity. In the first, the humans are placed in untamed wildness; in the second, a bounded garden. In the first, the humans come at the end of God's creating; in the second, the human comes at the beginning. In the first, there is unconditional abundant blessing; in the second, a task, with abundance to follow. In the first, there are rules about what the humans can do; in the second, what they can't. The stories highlight two vying animating principles that inform the human relationship to the world. In the wilderness, the call is for survivability, allowing the humans to take what is available from earth's bounty to satisfy their own needs and desires. In the garden, the call is for sustainability, with the human tending to the world's potential, serving the needs of all.

Throughout much of Western history, the immediate needs of survivability (and too often the urgings of gluttony) overwhelmed the pursuit of sustainability. And no wonder, given the precariousness of life for most of human existence. It was all we could do to fight back hunger, drought, cold, disease, infections, and the terrors of the night. Yet the more our technology and capacity grew, the more our sense of hubris grew, the more our desire for consumption grew. Today we know that to consume too much of earth's resources harms both our bodies and the body of the earth, which means we know that our survivability depends on the earth's sustainability.

It is by embedding the ethic of Genesis 1 in the ethic of Genesis 2 that we can learn to craft a right relationship with the earth.

Genesis 1: The World in Service to Humanity

Genesis 1 describes how the world came into being, step by step, layer by layer, over six cosmic days: Light vanquished darkness, the heavens were separated from the earth, dry land emerged from the waters, vegetation from the dry land, life forms of all kinds were fashioned according to their domains, and ultimately the emergence of humankind. At each step of the way, God examined what had been made, assessed it, and declared it good. Creation was the uncontested, intentional work of God, each step building up to the next, the whole serving as the stage upon which the drama of life would unfold.

Once the inaugural act of Creation was complete, the next question was how to keep it going forward. The answer was implanted in the primordial design of Genesis 1: life's capacity for self-renewal, self-regeneration. "And God said, 'Let the earth sprout vegetation: seed-bearing plants, fruit trees of every kind on earth that bear fruit with the seed in it.' And it was so.…And God blessed [the animals], saying, 'Be fertile and increase, fill the waters in the seas, and let the birds increase on the earth'" (Genesis 1:11, 1:22). One generation bore within it the life force of the next. Every new being was the parent to the next. Such powers of renewal were shared by plants and animals alike.

Life, from inception, was designed to yield more life. Each living entity in every generation was both creation and creator bearing the capacity to bring forth life after it, infinitely extending the realm of existence. God was not a constant fabricator, endlessly churning out individual pieces from scratch. God's prototype itself had the capacity to bring forth life in its image.

The implicit promise of Genesis 1 is that life—ever renewing—is the chosen, cherished expression of the Divine. The world was set on a path, programmed to go on endlessly, one generation effortlessly cascading into the next. Nachmanides, the famed thirteenth-century biblical commentator, captured this sense, writing, "God created species in the world... giving them the power to give birth so that those species should continue forever, as long as God wished the world to continue."[45] Life, in short, was created and designed to bring forth more life.

Vegetation and animals, place and resident, were well matched, each being what and where they were meant to be. The fish were made for the waters, and the waters were made for the fish. The birds were made for the skies, and the skies were made for the birds. The trees combined to make forests, the grass to make fields. All filled the earth so that it brimmed with life, the entire world humming along in a harmonious whole, each reproducing after its kind. Then and only then, when everything was ready in this first story of Creation, did God form humans.

"And God created the human in the divine image, creating the human in the image of God—male and female creating them" (Genesis 1:27). This last act of Creation was a magnitude apart from all the rest. These creatures could transcend domains. They possessed a touch of divinity. They were given a distinct role superior to all the rest of the animals. God blessed this male

One of the finest explanations of what it means to be made in the image of God was offered by the twentieth-century rabbi, scholar, and philosopher Joseph B. Soloveitchik. Being made in the image of God, he says, refers to humanity's "inner charismatic endowment as creative being." That is, the divine aspect of our humanity enables us to see beyond the world as it is and imagine it as it can yet become. This, says Soloveitchik, is the passion and brilliance of what he calls Adam the first, the human as created in Genesis 1.[46]

Lest humans become too prideful, a midrash (rabbinic commentary on biblical texts), sought to rein them in. Rashi (eleventh-century master commentator) wrote that though the text says the first humans were blessed by God with the ability to subdue (*yirdu*) the animals (Genesis 1:26), the word *yirdu* could be interpreted in two ways, as meaning either subduing or being subdued. That is, the human relationship to the animals was conditional. If the humans merited it, the animals would be good partners with them. But if the humans did not merit it, the animals would rule over them. Rashi does not elaborate on what it takes to merit a right relationship with the animals. We seem to still be discovering that.

and female, and said to them, "Be fertile and increase, fill the earth[47] and master it; and rule the fish of the sea, the birds of the sky, and all the living things that creep on earth" (Genesis 1:28).

The world in Genesis 1 was primed for humanity. The first five and a half days of Creation, good as they were in their own right, were nonetheless preparation for the best yet to come. Humanity was the pinnacle. After that, God stopped creating, assessed the whole, and declared it very good.

The world the first humans entered was one of untamed wildness. It is no wonder, then, that they were given permission, even a mandate, to master the environment, to use and subdue its bounty as needed. Survival and regeneration were the order of the whole enterprise.[48] In the wilderness, as reflected in Genesis 1, the first humans had only their wits to protect and sustain them. And in reality, too, humans over the first millennia of their existence were few. Their ability to manipulate their environment was naïve compared to the immense power of nature. Their footprint was small and their vulnerability great. The ingenuity and technology that would be the hallmarks of our species and the birthing ground of civilization were still in their nascent stages. The first humans in both the Bible and history had little protection from predators, disease, hunger, cold, heat, storms, darkness.

Genesis 1 seems to have tapped into humanity's primal anxiety and responded with the greatest reassurance found in its spiritual arsenal: God's blessing of power and control. Humans were made in the image of God and were gifted the capacity—and the right—to rule, explore, be curious, experiment, and use all manner of Creation and the gifts of inspiration with which they were endowed to enable them to flourish in this God-given world.

This was the initial expression, the first iteration of the call to create a habitable world, the mitzvah of *yishuv ha'olam*. Genesis 1 sees human survivability, staying alive, as the way to habitability. Sustainability was not its concern. The earth, it was thought, could manage quite well whatever early humans threw at it. Land, air, the waters, vegetation, animals all seemed boundless, indestructible, wholly giving, wholly forgiving. Humanity could extract what we needed from the earth, use what we wished, discard what was left. The immensity of the world, the unfathomably rigorous rhythms of its systems, reinforced the belief that earth was indefatigable, regenerative, impervious to our deeds and its gifts, there for the taking.

We are not so far from this early mindset as we might imagine. With the increase of antibiotic-resistant bacteria, the recent and yet-to-come worldwide deadly pandemics, droughts, floods, killer heat brought on by the advent of climate change, and, some might even add, the advent of artificial intelligence, our vulnerability is increasing.

Such was the Genesis 1–sanctioned way of engaging with the world. In the face of our vulnerability, at the mercy of predators and disease, of plagues and infections, of famine, fear, and loss, who would not want the right and capacity to subdue nature and have it do our bidding? Who would not want to be able to vanquish illness, hardship, and death? It is the security of modernity that has allowed us to tame the earth's wildness in the human imagination. Back in the day, when death and hunger lurked around every corner and disease could be transmitted by the slightest nick or sip of water, knowing that one's founding myth validated, encouraged, and enabled human mastery of nature could be profoundly reassuring.

This did not mean that humanity, per the story, was given totally unrestrained access to the world. Humans might be in the image of God, but they were not God. The first humans were not able to pursue their appetites

Over the centuries, eating meat on Shabbat was seen as a way of celebrating and honoring the holiness of the seventh day. But given the juxtaposition of texts here in Genesis, where the utility but not consumption of animals is stressed, it seems appropriate that we reexamine the cultural role that meat-eating plays in our lives today, especially since we now know that animal agriculture contributes up to 20 percent of annual greenhouse gas emissions, exacerbating climate change, as well as causing 41 percent of global deforestation,[50] plus dead zones in coastal areas, water waste (it takes 1,847 gallons to create a pound of beef versus 39 gallons to create a pound of vegetables),[51] and land waste. Reducing our meat consumption could redirect resources that could more efficiently grow food directly for human consumption, thus reducing the negative impact of much of our food industry.

boundlessly. Animals—in Genesis 1—were to be subdued and mastered, not eaten.[49] And the celebration of Shabbat, a day of cosmic rest immediately following humankind's creation, reminded the humans to whom the earth truly belongs and signaled to them that the mastery they were given has its limits.

Still, coupling humanity's precarious position with earth's regenerative capacity, the big takeaway from Genesis 1 is humanity's ability and privilege to exploit the gifts of the world. It seemed reasonable then, to so many over the centuries, that humankind was justified in using the earth as we saw fit.

So we did, for thousands of years. We tamed the rivers and felled the trees; distilled potions from the plants to reduce our fevers and ease our pains. We pressed animals into service of all kinds and built cities, roads, and museums. We planted and harvested and gathered in, altering the land and affecting the animals, the flora, the watercourses as we went. We were fruitful and multiplied and burst the bounds of God's blessing. In short, Genesis 1 was seen as justifying the view that humans are the rightful primary beneficiaries of the goodness of the earth, seeking to flourish in most any way we could in an eternally self-regenerating world. And we lived fully into those blessings.

But the innocence of that vision—given how easily use slips into abuse—can no longer guide us. Its pursuit, we now know, has led us headlong into a global nightmare.[52] Though as individuals we are still small and at the mercy of nature's power, collectively we are neither few nor weak nor insignificant. Our footprint is larger than it has ever been. Over the last 150 years, we have profoundly disrupted the world's operating systems in ways that are irreparable in any meaningful human time frame. We are changing the character of the atmosphere and the fertility of the oceans, the pattern of rainfall and the contours of the land.

Genesis 1 is a story that offers counsel and comfort within a consciousness of high human vulnerability and low human capacity to disrupt the world. Things are different now. Genesis 1s survivability premise has been found wanting. The limits of humanity's ability to flourish while subduing, filling, and harming the earth have been reached, even overrun. Humanity is far outpacing the earth's capacity to support us. To pursue unchanged the blessing we believe we were given to master the world as we have done in the past will bring devastation. What was once seen as the key to survival now threatens us. We need a new story, a new guiding vision, centering on the whole of Creation, not just humanity, calling us to care for nature as a system so that we may enable it to continue to care for us. Fortunately, we have that story… in the very next verses of the Torah: Genesis 2.

Genesis 2: Humanity in Service to the World

Genesis 2 seems to start all over again, opening with a description of the earth radically different from Genesis 1. "Such is the story of heaven and earth when they were created. When God made earth and heaven, no shrub of the field was yet upon the earth and the grass of the field had not yet grown, since God had not caused rain to fall upon the earth, and there was no human to till it" (Genesis 2:4–5). Where an abundance of water and the need to constrain it were the challenges in Genesis 1, the scarcity of water and the need to coax it into being, so it flowed from the ground and fell from the skies, were the challenges of Genesis

Humus is earth, soil (not to be confused with hummus, the Mediterranean dish made with chickpeas and tahini). And not just any soil, but soil that is rich in nutrients, soil that is arable, able to support lots of plant life, including those used in human agriculture. This linguistic connection between humanity and humus was made prominent by Rabbi Arthur Waskow, echoing the connection of *adam* and *adamah* in Hebrew.

2. Where Genesis 1 opens with an explosion of light, Genesis 2 opens onto a colorless world, dry, bleak, and barren. For two essential ingredients are missing: rain and humanity. Unlike in Genesis 1, where God tamed the wild, ubiquitous profusion of waters and brought into being all that was needed to ensure the well-being of humankind, in Genesis 2, God loosed the waters trapped in the ground, allowing them to flow and water the earth. Then God fashioned the human and began a garden, anticipating that both the waters and the human would contribute to its well-being.

The world of Genesis 2, unlike Genesis 1, was not a smorgasbord laid out in its fullness by God for humankind's taking. It was not a world that could thrive independent of the vigilant, skilled input of humankind. Genesis 2 presents a world in which life depends on the prowess and attentiveness and efforts of humankind.

This is a radically new vision of earth's creation and human vocation. If Genesis 1 is an ego-driven narrative, placing humans in the center and nature in service to them, Genesis 2 is an eco-driven narrative, placing nature in the center and humans in service to it. Instead in Genesis 1, humans were the unconditional beneficiaries of a pre-fabricated delight laid out for our use and hegemony. In Genesis 2, the bounty of the earth—the vegetation, the animals, the regeneration to come—would appear only with the nurturing work of humanity and the nourishing water given by God.

Genesis 2 positions humans as helpers, God's partners, not independent sovereigns. It recognizes our capacity and ingenuity and crafts a way for us to care for the earth both wisely and well, not just for ourselves but for the good of the whole. We are seen as part of—not apart from—earth's infrastructure designed to bring forth a verdant world. Our task is to care for the greater community of life and to understand that that is the best way to care for ourselves. To strengthen this affinity with the object of our work, the story tells us that we are fashioned from the stuff of earth: *adam*, the human, is made from *adamah*, the earth. Humanity from humus. Not by the incorporeal speech of the Divine, as in Genesis 1, but from the mud and soil of the fertile earth we depend on.

Genesis Rabbah (8:1) tells us that the first Human was in fact created as two, *du partzufim*, one being binding together both male and female. This is a rabbinic attempt to reconcile the differing descriptions of the creation of humankind, *adam*, found in Genesis 2 and Genesis 1.

There is a subtle word shift in the opening verses of Genesis 2 that signals that we are in the presence of a different vision of the relationship between earth and humankind. When speaking of the earth, Genesis 1 prefers the word *eretz*, meaning "land, natural terrain, earth, the ground beneath our feet," the stuff—as my father used to say—that they aren't making any more of. *Eretz* is nature, Creation in the raw. Genesis 2, however, prefers words like *sadeh*, "field"; *adamah*, "arable soil"; and *gan*, "garden"—words that speak of cultivated areas contoured, planted, and worked by human hands. In Genesis 1, the earth seems to bring forth food on its own, feeding humans and animals as a matter of course. Genesis 2 requires the partnership of humans, soil, rain, and God to bring forth foods from the earth.

And to sharpen the human's role here on earth, Genesis 2 explicitly states, "God settled the Human in the garden of Eden, *l'ovdah ul'shomrah* [to work it and protect it]" (Genesis 2:15)—to till it and tend it; to serve and preserve it. Our task, in short, is to nurture the land, enabling it to be all it can be and do all it can do.

Many commentators have noted the multiple resonances of these two Hebrew verbs, *l'ovdah ul'shomrah*. The word for "work it" (*l'ovdah*) shares its root with the word for sacred service (*avodah*), evoking the Temple

Genesis 2 presents humans as life-giving as the rain. We are depicted as indispensable to the emergence and productivity of a thriving earth as the waters from heaven or the streams from below. But there is a caution there as well. For as water can become a curse, so can humanity. Rain inappropriately apportioned or ill-timed, can be destructive, even deadly.[53] As can we. Consume too much and we drown the earth in our waste and our pollutants, our ill-conceived and misplaced construction, our disproportionate demands of earth's resources. And in the shadow of our consumption, we starve other creatures by taking, destroying, diverting, denying what they need. Our task is to be like the life-giving rain.

cult, with its songs, sacrifices, reverence, and awe. To work the earth, then, is to perform a sacred service, to tend it in reverence, awe, even love. As the farmer/poet/philosopher Wendell Berry writes, "We have the world to live in on the condition that we will take good care of it. And to take good care of it, we have to know it. And to know it and to be willing to take care of it, we have to love it."[54] The agricultural laws of the Torah, the grains and fruits the Israelites were commanded to give as tithes, the sacrifices that were to be taken to the Temple, all reinforce this vision of nature as a sacred medium with which our ancestors partnered with and expressed love of God. That is what we were created to do, the Torah tells us—to be the caretakers, the gardeners, of God's sacred world.

The word "protect it" (*l'shomrah*) yields rich multiple associations as well. Its root can mean "to keep, to practice, to observe," as in observing the Sabbath and other mitzvot, as well as guarding and protecting something owned by someone else (in this case, God) that is left in our care. We are not just the gardeners, then, but the "guardeners" of God's earth—the ones who work it, know it, protect it, nurture it, come to love it, and allow it to burst forth with life under our care.

But the text then conjures up an interesting question: From whom or from what at this dewy moment of Creation are we protecting the garden? Surely the Creator was not going to destroy this newly wrought paradise.[55] (At least not now, not yet. That would wait for the debauched generation of Noah.) Neither would, or could, the animals. The only source of danger to the garden was the human.[56] In the protector lay the seeds of the destroyer. Genesis 2 knew that the privilege of use can lead to the excess of misuse, even abuse and plunder. Given the right conditions, the ones charged with caring for Creation could become the source of its destruction. Hence the need to make the task of working the earth a sacred charge, balanced with the responsibility of protecting it.

There is yet another way Genesis 2 differs from Genesis 1. Genesis 2 teaches us about restraint and the reality of limits: "And the Sovereign God commanded the human, saying, 'You may eat from every tree of the garden, save for the tree of the knowledge of good and evil; do not eat from it, for if you do, on that very day, you shall die'" (Genesis 2:16–17). All life has its limits, all being has boundaries. Some may be tugged and stretched a bit here, pushed a bit there, but none may be crossed. We humans (curious and adventurous creatures that we are) tend to learn the bounds of many limits only after we have violated them. So it was with Adam and Eve, who ate the fruit from the forbidden tree and unleashed irreparable havoc, exiling themselves from the generous garden, only to enter a miser's lot of thistles and thorns. We don't have the luxury of making that mistake when the very existence of life on earth is at stake.

Genesis 2 warns us of that and shows us the price of overreach, of believing that we know more than we do, or of being reckless with capacities that might outstrip our ability to control them. Our role as *guardeners* requires us to explore, be creative, and improve both ourselves and the world around us. But it also cautions us to do so prudently.

Still, caution should not stymie progress. Genesis 2 offers a warning, not a proscription. Through our curiosity we know many impressive things, have created abundant blessings in this world, and are uncovering some of the deepest mysteries of life. But we must be cautious about whether, when, and how we can truly, sufficiently, safely assess benefit from harm, good from evil, development from decay, desire from need. For it is the seductive nature of progress to initially reveal only part of her truth. And as Genesis 2 tells us, what we seek to know can be deadly.

The story of Genesis 2, therefore, proclaims a different message from that of Genesis 1, and its narrative yields a different theological charge. Creation here was not a linear, layered hierarchy, but a grand weave of interdependence. Caring for self was bound up with caring for the earth. "Guardener" was a defining component of human identity. Genesis 2 places humanity in a garden, not a wilderness. Gardens are a symbiosis of nature and humanity. They require the diligence of planning and planting, the laying out of boundaries. They need tending, watering, patience, and more. And they require the gifts of nature: rain, pollinators, climate, sun, soil, time. The glory of earth's bounty is born of this courtship of humans and nature.[57]

Survivability amid Sustainability

Perhaps because of Genesis 2's placement (being second), or perhaps because of the lyricism and boundlessness of Genesis 1, or perhaps because Shabbat and abundance offer a happier ending than expulsion and scarcity, Genesis 2's message of how to care for the earth has largely been overshadowed. But the earth can no longer abide our indiscretions. And our misdeeds are no longer forgivable. For we know that survivability requires sustainability, that the health of the earth determines the well-being of humanity.

We see this truth all around us, not only in the material resources and ecosystem services we rely on each day, but also in the spiritual nurturing we receive from the earth. During times of stress, we seek solace and comfort in the woods, the mountains, the seashores, the green spaces. Our public parks are increasingly strained by our desire to find in them refuge from the stress of our daily lives. We go to them as a form of meditation, a spiritual medication. The Japanese have a word for it: *shinrin-yoku*, "forest bathing." Encounters with nature can soothe and restore us. Even just seeing nature through a window or on a screen aids patients recovering from surgery.

We need the bounty of this good earth for so many reasons. Yet the fires, the flooding, the droughts, the melting of the sea ice, the rising of the waters, the expanding range of disease, the loss of species, the mutations of new viruses, the instability of the atmosphere, the displacement of millions of people due to land that can no longer support them tell us that we have betrayed ourselves and burst the bounds of earth's limits.

The stories of Genesis 1 and Genesis 2 both seek *yishuv ha'olam*, the reassurance of a habitable world. They both tell a story of the relationship between humans and nature—in vastly divergent ways. As we today stand on the precipice of a sixth grand extinction, sliding into a climate we have never before known, we are called to move from a Genesis 1 vision of *yishuv ha'olam* to a Genesis 2 vision so we may design ways that renew, preserve, and reuse earth's resources in an endless cycle of ever-emerging Creation.[58]

This is a personal task, a communal task, a political task, and a marketplace task. We are all in this together. Personal habits that align with this goal are essential. But personal practices alone are not enough. We

A mass extinction is the rapid, widespread loss of life beyond the normal ebb and flow of evolution. The most well-known mass extinction is the Cretaceous-Paleogene extinction, which occurred sixty-six million years ago, caused, it is believed, by an asteroid impact. Seventy-five percent of all species on earth perished. Extinction, of course, means gone forever. Not for a little while; not even for a long while. But forever. Extinctions alter all the rest of history, for all that could yet have come from that life is now also lost. The current event, called the Holocene extinction, includes the loss of birds, insects, land animals, sea animals, plants. It is different from all the other extinctions, for this alone is caused by us. It can be stopped by us as well.

Rabbi Joseph ibn Kaspi, made the same point: "In our pride we foolishly imagine that there is no kinship between us and the rest of the animal world, how much less with plants and minerals. To eradicate this foolish notion God gave us certain precepts, some concerning minerals, others vegetable, others animal, and others human. Above all we are bidden to be compassionate." To humans, of course, but to all the rest of Creation as well.[59]

cannot recycle if most of our plastics are not recyclable. We cannot properly compost if there is no municipal food pickup. We cannot purchase sustainable products if they do not exist or are beyond our means. In the early twentieth century, cities developed better ways to manage water, waste, food, transportation. Today, we must develop earth-friendly, energy-efficient ways to manage our production and consumption—ways that we all use as a matter of course.

Modernity has long ago left behind Genesis 1's world of easy blessings, when we dared to imagine we could focus on our own well-being without caring for the well-being of the world. We must move into a Genesis 2 understanding where we know earth and humanity are mutually dependent, the one leaning on the goodness of the other.

The poet Mary Oliver once wrote "that there exist a thousand unbreakable links between each of us and everything else, and that our dignity and our chances are one. . . . We are each other's destiny."[60] Many cultures, including Judaism, have long known this truth and sought to live well within earth's bounds and shelter. Modernity must too. For when we do, we will become the agents that ensure the fullness of God's blessing spoken to Noah after the flood: "So long as the earth endures, seedtime and harvest, cold and heat, summer and winter, day and night shall not cease" (Genesis 8:22).[61]

Charting Genesis 1 and Genesis 2

This chart seeks to lay out some of the more salient differences between the Creation stories in Genesis 1 and Genesis 2, showing the difference between an ethic of survivability and an ethic of sustainability.

Genesis 1 Ethic of Survivability	**Genesis 2** Ethic of Sustainability
World is wilderness	World is garden
Domination and control	Work/labor & preserve/protect
Earth in service to humanity	Humanity and earth in partnership
Earth is self-renewing	Earth requires “inputs” of water and humanity
Humans are “godly”	Humans are “humus”
Human relationship to earth: extractive	Human relationship to earth: generative
Consumption (almost) limitless	Consumption limited
No concept of exile or the earth’s destruction	Concept of exile and destruction
Nature as commodity	Nature as community
Ego-centric	Eco-centric

CHAPTER FOUR

Holy Sparks—Divining Nature

The Baal Shem Tov, the fabled eighteenth-century founder of the modern Chasidic movement, taught, "It is a great principle that holy sparks dwell in all matter of the world. Nothing is void of sparks, even trees and stones."[62]

To believe this is to believe that all the world is enchanted, that it is infused with an essence pointing beyond the quotidian, deepening the experience of everyday life. Knowing so, saying so, is both good for the world and good for us. Good for us, for when we live in deep relationship with the land, we infuse it with stories, build rituals around it, weave bonds of belonging between us and place. We create safe anchorage for our souls from which we can set sail to explore the world. And good for the world, for when we are transported by nature's everyday enchantment, what was object becomes subject, what was inert becomes animate and we are moved to protect it. Preserving it becomes a moral act, prompted by love, not a utilitarian act prompted by need (if we are moved to protect it at all).

From the earliest days of the Bible to the sacred rituals of today, Judaism has acknowledged and celebrated the animating force that inhabits the world around us. Abraham planted a tamarisk in Beer Sheva and invoked God's name El Olam, "the God of all Creation." Jacob encountered God—or perhaps only God's angels—that starry night long ago, called that place Beth El, "the house of God," and marked it with a stone. The prophets spoke in metaphors of fields and mountains and trees and fruits and vines and wind and water to conjure the greatness of their sense of God, matching the outer landscape of the world to the inner landscape of their souls. For two thousand years, Jewish liturgy has urged us to say blessings upon witnessing a litany of natural events: lightning,

As a verb, to "divine"[63] is to use intuition or insight to discover or locate something that is otherwise hidden. As an adjective or noun (the more common usages of the word), it refers to a connection with God. I use it here in both senses. I imagine the ancestors landed upon their views of nature through intuition, or inspiration. And in doing so they connected what they discovered in this physical world to the supreme Divine.

thunder, the first blossoms of spring, the cycles of the moon.[64] We celebrate the seasons of the harvest, offer prayers for rain and dew. On Sukkot, the fall pilgrimage and harvest festival, we gather palm, willow, myrtle, and citron and wave them together. We sit in temporary dwellings covered by branches. An early rabbinic tradition even teaches that each blade of grass has an angel assigned to it, urging it to grow.[65]

The fear of animism often stultified Judaism's celebration of the sacredness within nature. The Bible itself repeatedly warns against worshipping gods and goddesses associated with trees or other elements of nature. Yet such fears could not undo the innate link that binds together heaven and earth. The Jews are a people whose first encounter with God was a call to the land. Centuries later, the Jews were forged into a nation through desert wanderings where they were freed from slavery to return to their land. Throughout the biblical period, God and the Israelites communicated through the medium of the land, that is, through the harvests that the land brought forth (God's gift to the Israelites) and the sacrifices offered at the Temple (the Israelites' gifts to God). Throughout the ages, the Jewish people found sacred meaning in "word and world alike."[66] We are, after all, the *am ha'aretz*, "the people of the land," as well as *am hasefer*, "the people of the book."

These efforts allowed Judaism over the centuries to masterfully blend what could otherwise be seen as two irreconcilable domains: spirit and matter, the celestial and the earthy, the eternal and the fleeting, the transcendent and the immanent, the infinite and the finite. In these ways, God—the holy and wholly other—was also made present here and now. Nature was made hallow as an expression of the Divine. Our relationship to nature was embedded in our sacred practices.

And yet, much of Jewish tradition remained fearful, or perhaps at least wary, of dwelling too much on a direct appreciation of nature. A teaching of Rabbi Jacob (or some say Rabbi Shimon) in *Pirkei Avot* (the two-thousand-year-old collection of rabbinic teachings and aphorisms) became the epitome and justification of such fear: "If one is studying while walking on the road and interrupts his study and says, 'How fine is this tree!' [or] 'How fine is this newly ploughed field!' it is as if he condemned himself."[67]

Many rabbinic authorities over the centuries have struggled with this teaching. Unhappy with its message, yet uneasy contradicting it out of respect for early traditions, they worked to find ways to blunt its harsh conclusion.[68] Still, echoes of this sentiment pervaded much of mainstream rabbinic thought, dulling formal recognition and celebration of nature in Judaism. And yet, the celebration of nature persisted. No less a leading light than the popular nineteenth-century rabbi, Joseph Chaim Caro (not to be confused with the author of the *Shulchan Aruch*) wrote in his commentary to *Ethics of the Fathers* how nature can be a pathway to God:

> There are certain people who know the Creator through the world's amazing creations, as it says: "The heavens declare your glory, God" [Psalm 19]. And it is not only in the high heavens that one can discern the actions of the Creator, but also in the seed in the ground, and in the fruit of the tree that grows in the field, in the lowly mosses that can be seen growing on the walls, all the way to the mighty, lofty cedars

> of Lebanon, because all these express the might of God and God's wonders.[69]

There is no one through line in Judaism that encompasses a unified approach to nature. Trends of wariness and celebration weave themselves in and out of centuries. But, through a parade of texts culled from the ages, what this chapter seeks to show is that there has always been an affinity in Judaism to see nature as conduit to and emblem of the sacred.

Seeing the Sacred All Around Us

None of us needs to be expert in some secret spiritual discipline or privy to the mantric ways of the mystics to sense the sacred all around us. We can feel it in the everyday, when we are captivated by the roseate colors of dawn that conjure the hues of the first creation. We can feel it in our breath that flows in and out, animating us each and every moment. We can see it in the insistent vitality of life that can be found even in places like Chernobyl, which has become Europe's third largest nature preserve, "with lynx, bison, deer and other animals roaming through thick forests,"[70] and through the grasses that insistently punch through cracks in urban sidewalks. We can see it in the two-thousand-year-old date seed found at Masada that Drs. Sarah Sallon and Elaine Solowey dared to plant in 2005; they rooted it, and it grew and is flourishing to this very day. We can see it in the Callery pear tree that survived 9/11; each year, three communities that experienced trauma receive seedlings from this survivor tree, branches of hope symbolizing the tenacity of being. All around us are thousands of other examples of the persistent audacity of nature.

Eighty percent of the US population lives in metropolitan areas where noise, hard surfaces, tall buildings, and bright lights blunt our raw experience of nature. Even more, such sprawling urbanism often denatures whatever parts of earth's greenness we do see, obscuring the sparks. Our vast consumerism and acts of commodification cause us to call much of the world not "nature" but "natural resources," as if earth were all one large storeroom, ever at our disposal.

Yet we are urged to see beyond the purely utilitarian and discover more in both word and world. That is why so many letters of the Torah have *tagin*, crowns, ornamentation, on top of them. The words would be perfectly understandable without them. They are there, we are told, as inspiration and hints to more expansive meanings and additional dimensions beyond the straightforward interpretation of the text.[71]

Scribes often decorate letters of Torah with crowns, hinting at deeper and more expansive meanings to the text.

Jewish liturgy reinforces the celebration of these expansive meanings. It sees sunrise as more than just sunrise. Each is a reflection of the light of the first dawn. Sunset is more than just sunset. It is part of a nightly celestial parade. Our morning prayers tell us, "God compassionately illumines the earth and all that dwell on it, continually renewing the work of Creation." Our evening prayers tell us, "Blessed are You, God, who brings on evenings with a word, and with wisdom opens the gates (of heaven). You wisely provide for the cycles of times and the flow of the seasons. You array the stars according to their assigned watch, creating day and night, rolling the light away from the darkness, and darkness away from the light, distinguishing day from night." God, our tradition imagines, is a conductor, cuing the rhythms of the skies like a maestro cuing an orchestra.

Magic Eye

Serendipitously, in the early 1990s, a series of books came along that offered to even the most skeptical among us a way into the mystical vision of our ancestors. (That was not their intent. The books were marketed merely for fun, and profit. Still, the novelty they offered opened a portal to a hidden world, captivating the imagination of millions and frustrating almost as many.) The series was called *Magic Eye*. It consisted of pages and pages filled with shapes and colors in repeating patterns arrayed across the page. Nothing more. No words. But, these otherwise unremarkable visuals concealed a remarkable secret: a third dimension that lay beneath, somehow within, the two-dimensional images we could see. The trick was to find it.

That was no easy task. Instead of focusing on the surface of the page, readers had to focus on what was hidden and beyond. They could only do that by looking not *at* but *through* the page to encounter something new, something that would open itself to them—though they knew not what that was. And to do that, first they had to believe, or at least hope, that something was indeed there, that their effort wouldn't be futile, that their trust was not misplaced. In short, they had to take a leap of faith.

Magic Eye calls us to believe that what we see on the surface might not be all there is, that our vision of the world is limited only by how we choose to see—and that it is up to us to seek what hides in the depths and find it.

The *Zohar* (the premier book of Jewish mysticism) deepens this vision of God active in the here and now. Every day, it teaches, "a ray of light [from the divine realm] shines into the world, keeping everything alive; with that ray God feeds the world."[72] Life and its endurance depend on that primordial ray of light that burst on the scene on the very first day of Creation. It was a light so radiant, so powerful, that it filled up the universe from one end to the other. This light of the first day was of a different nature than the light of the sun, which was created on the fourth day. This life-giving light, if let loose in its fullness, would (paradoxically) have overwhelmed the material world. So to accommodate the fragility of the matter to come, God withdrew the fullness of this primordial light and each day sends to earth instead one ray, a jet from the luminous vaults of heaven. It is that light, that ray, says the *Zohar*, that keeps the world alive.

Abraham Joshua Heschel, the twentieth-century mystic and philosopher who spoke so often in a prophetic key, felt the presence of this sacred light and urged us to as well. But he feared it was being obscured by our modern habits. Living as we do, he taught, inside cathedrals of space, bathed constantly in artificial light and year-round temperature control, with nature something that we walk through rather than dwell in, we miss the intimate connection with the cycles of our days. Our experience of nature, he believed, is being blunted, causing us to lose touch with the larger essence that infuses the world around us. "Away from the immense, cloistered in our own concepts, we may scorn and revile everything."[73]

Yet, he wrote, there is a ready answer. Stand outside (find a suitably dark place!), and look up. "No one

Late one Shabbat afternoon, my granddaughter and I sat talking in a dimly lit room (we don't turn lights on or off on Shabbat, and this room was lit only by a modest reading light). Outside, dusk was gathering slowly. At one point, my granddaughter turned to the window and said with a touch of awe, "Look! It's getting dark!" As if to say: "Wait! Is this what happens every day?" With our artificial lights, we rarely notice the dimming of the day's sunlight. We go from light of day to the bright of night and forget that we are riding on a slowly turning globe suspended in the unfathomably deep and dark eternity of space.

can sneer at the stars, mock the dawn, ridicule the outburst of spring or scoff at the totality of being. Standing between heaven and earth, we are silenced."[74] Few can behold the universe, its splendor, its immensity and complexity, he argues, and remain unmoved. To peer into the depths of time and see across billions of miles of space, to witness the Perseids every August, or here on earth to watch a tendril erupt from a seed, to welcome the birds returning to the nests outside your window every spring, is to know wonder. And more, it is to realize that we are not simply witnesses to this universe, but part of its body, its destiny, its drama.

To all that, Heschel teaches, our response should be awe. "Awe," he writes, "is an awareness of the *transcendent worth of the universe*." It is to realize the magnificence of what we take for granted. If the stars came out once every thousand years, who would be able to sleep that night? If there were but one tree on earth, who would not make pilgrimage to see it? The constancy of these miracles tends to blunt our appreciation of them. Yet we need only to open ourselves again to nature's awe, to allow ourselves to be in thrall to her wonders. In so doing, we will transcend, even for a moment, the limits of our bounded sense of self.[75]

In a cartoon that opens one of his collections, Bill Watterson, the creator of the incomparable comic strip *Calvin and Hobbes*, shows Calvin (a mischievous six-year-old) standing shoulder deep in a hole he just dug, shovel in hand. Hobbes (his stuffed tiger) is leaning over him straining to see into the hole. Hobbes asks, "What have you found?" Calvin answers, "A few dirty rocks, a weird root, and some disgusting grubs." Hobbes, amazed at Calvin's good luck, responds, "On your first try?" Calvin, exuberant, pronounces, "There's treasure everywhere."

Indeed it is, if we know how to look for it.

Expressions of awe reside not only in the writings of mystics, scientists, and poets. Rabbi Joseph Soloveitchik, who spoke of the divine "inner charismatic endowment" of human beings, also urges us to find holy sparks in the ways of nature. We can look for the image of God, he writes, "in every beam of light, every bud and blossom, in the morning breeze and the stillness of a starlit evening."[76]

So too the twelfth-century philosopher and ethicist Bachya ibn Pakuda taught that it is not only through mystical association but also through experiencing the world and all its ways that we can encounter the Divine. "It is our duty to explore the work of Creation, and discover therein the proofs of Divine Wisdom... in its form and shapes and purpose; the glorious spirituality in its materiality... all this points to the Creator as clearly as a piece of work points to the workman or a house to its builder."[77]

Such teachers not only allow us to discover the divine in the midst of the mundane, they encourage us to do so.

A Chasidic teaching as retold by Rabbi Daniel Swartz: If you were to walk through the woods and a spring appeared just when you were thirsty, you'd call it a miracle. If it were to happen again, you'd call it a coincidence. But if that spring was there all the time, you might not notice it at all—but wouldn't that be the most wondrous of all?

The Bible's Enchantment with the Land

The authors of the Bible lived in deep relationship with the land. They knew its seasons and its moods, its needs, its blessings, and its curses. They spoke of its destructive powers in their stories of the plagues in Egypt and of its beneficence in their poetry of the Promised Land, "a land flowing with streams and springs and fountains issuing from plain and hill; a land of wheat and barley, of vines, figs, and pomegranates, a land of

Richard Feynman: On Creation's Majesty

Science is no stranger to such amazement. Discovering the intricacies of life often feed that sense of joy and wonder. Richard Feynman, the iconoclastic twentieth-century theoretical physicist, continually spoke of Creation's majesty. In this musing, [UNTITLED ODE TO THE WONDER OF LIFE], he writes:

> I stand at the seashore, alone, and start to think. There are the rushing waves... mountains of molecules, each stupidly minding its own business... trillions apart... yet forming white surf in unison....
>
> Deep in the sea, all molecules repeat the patterns of one another till complex new ones are formed. They make others like themselves... and a new dance starts.
>
> Growing in size and complexity... living things, masses of atoms, DNA, protein... dancing a pattern ever more intricate.
>
> Out of the cradle onto the dry land... here it is standing... atoms with consciousness... matter with curiosity.
>
> Stands at the sea... wonders at wondering... I... a universe of atoms... an atom in the universe.

Richard Feynman first shared this poem while giving a speech in 1955 at the National Academy of Sciences. It later became the epilogue in his last book, a memoir entitled, *What Do You Care What Other People Think?*

olive trees and honey" (Deuteronomy 8:7–8). They knew the finicky personality of its rains and the stubbornness of the soil. And still they loved it.

Children in the biblical period (ca. 1000 BCE) were thought to have learned and chanted the ways of the seasons as we learn and chant our ABC's. Knowledge of the agricultural cycle and its synchrony with the months was invaluable in the training of future farmers. This calendar found in the ancient city of Gezer, twenty miles west of Jerusalem, might have been a classroom primer:

> Two months gathering
> Two months planting
> Two months late sowing
> One month cutting flax
> One month reaping barley
> One month reaping and measuring grain
> Two months pruning
> One month summer fruit[78]

For the biblical audience, nature was more than a place to live and work, even more than the source and substance of life. It was the meeting place of Israelites and God, the medium through which God and Israel communicated. God expressed pleasure or displeasure through nature. The land would experience rain or drought, abundance or dearth, peace or upheaval, all in response to the Israelites' behavior. Though the text vigorously distinguishes nature from God, for God transcends nature, still the earth was seen as an expression of the Divine. Earth was gift; God was the Giver. The workings of nature were infused with, and dependent on, the desires of the Divine.

This is why the land of Israel was and remains such a central part of the sacred story of the Jewish people. In the biblical period, the land was celebrated every day through practices of the Temple. The sacrifices, the main ritual activity of the Jewish people, were grains and fruits and animals—products of nature. These were received as gifts from God—bestowed on humanity through the medium of land, sun, rain, fertile soil, stable climate, the farmer's health, life's regenerativity—and returned to God through fire, the medium that mixes matter and air, heaven and earth. The sacrifices were called *korbanot*, evoking the sense of "drawing near"; that is, the sacrifices were designed to bring God and humans closer together. Whether they were the voluntary offerings of gratitude, the obligatory offerings of contrition, or the ongoing offerings determined by the sweep of the calendar and the order of the day, the purpose of the sacrifices was to reach across the divide and span the gulf between heaven and earth.

As a medium connecting the celestial and human domains, the land was perceived as more than land, even more than a connector. It was seen as a member of the sacred covenantal community, animated, even sentient; a moral entity, judging and responding to Israel's behavior, reflecting and executing God's designs, guarding God's values. The biblical imagination, in short, imparted agency to the land. The land demanded that the people treat it and one another well. Immorality offended it. Right behavior delighted it. Be good and it will bring forth blessings, the Torah proclaims: "The mountains will produce well-being for the people; the hills, the reward of justice" (Psalm 72:3). Mistreat it, or one another, and the land will spit you out. "By all those abhorrent things done by the people who were in the land before you, the land became defiled. So let not the land spew you out for defiling it, as it spewed out the nation that came before you" (Leviticus 18:27–28).[79]

The prophets and psalmists go beyond endowing nature with agency. They endow it with feelings and expressiveness as well. Nature becomes a rhapsodic chorus, an exuberant, vocal witness exalting God's grandeur, filling the earth with praise and hosannas. "The heavens declare the glory of God," reports the author of Psalm 19, "the skies proclaim God's handiwork" (verse 2). Psalm 96 calls nature forth: "Let the heavens rejoice and the earth exult; let the sea and all within it thunder, the fields and everything in them exult; then shall all the trees of the forest shout for joy" (verses 11–12).

As a character in the sacred story, the land not only reacted to Israel's disobediences, but was depicted as suffering with them, on account of them. "How long must the land be in mourning," Jeremiah asks, "and the grass of all the countryside dry up? Must beasts and birds perish, because of the evil of its inhabitants?" (Jeremiah 12:4). Hosea likewise laments, saying, "The earth is withered. Everything that dwells on it languishes. Beasts of the field and birds of the sky, even the fish of the sea perish" (Hosea 4:3), because humanity has become debased.

The land was seen as an active party, a third member to the divine/human covenant. As the biblical scholar Johannes Pederson daringly put it, "The earth has its nature, which makes itself felt and demands respect."[80] So it is that God makes a treaty not only with humanity but directly with the land. After seeing the devastation wrought by the flood, God says, "I have set My bow in the clouds, and it shall serve as a sign of the covenant be-

The prophet Joel speaks his prophecies through images of nature and agriculture, immediately familiar to his biblical audience. But he does one thing more. He speaks not only *about* the land and animals but *to* the land and animals.

Fear not, O soil,
rejoice and be glad;
For the Eternal has
wrought great deeds.

Fear not,
O beasts of the field,
For the pastures
in the wilderness
are clothed with grass.

The trees have borne
their fruit;
Fig tree and vine
have yielded
their strength.
(Joel 2:21–22)

tween Me and the earth" (Genesis 9:13). God would no longer flood the earth because of human misdeeds. If ever again the oceans overtake the shore and the rains flood the interior, if lives are lost and species die, that is on us.

And not only with floods. Earth's joys and sorrows, health or illness, depend on us. If we treat the land or one another badly, the Torah warns, the land will rebel. Morality spills over into the realm of materiality. How we treat people and how we treat the land more often than not go hand in hand.

So central was this teaching that the message was inserted into the daily liturgy, recited as the second paragraph of the Sh'ma. In this paragraph, taken from Deuteronomy 11, God uses the medium of the land to mete out judgment or blessings on the people. Every morning and evening those who pray the traditional liturgy say, "If, then, you obey the commandments that I enjoin upon you this day, loving the Eternal your God and serving [God] with all your heart and soul, I will grant rain for your land in its proper season, the early rain and the late. You shall gather in your new grain and wine and oil—I will also provide grass in the fields for your cattle—and thus you will eat your fill" (Deuteronomy 11:13–15). But misbehave, the Torah says, and violate the laws (whether ritual or ethical) that God set before you, and "the Eternal's anger will flare up against you, shutting up the skies so that there will be no rain and the ground will not yield its produce; and you will soon perish from the good land that the Eternal is assigning to you" (Deuteronomy 11:17).

This ancient text, whose theology of divine reward and punishment sits uncomfortably with many today, has had a renaissance of sorts. For read deeply, it tells us that it is not God but our actions that will bring blessings or curses upon the land. It teaches us that what we do, how we treat one another and how we treat the land, has consequences. The inability of the land to flourish is often the inevitable result of our injudicious use of the plow, pesticides, fertilizers, logging; the result of clear-cutting forests, removing mountaintops and dumping the debris into local streams, fracking deep in the earth and poisoning the ground water—and the unethical treatment of those who do the work.

How well we care for the land determines how well the land will care for us, which determines how well we care for one another. Across the span of history, abuse of the land almost always entailed the abuse of the other, whether through slave labor, imposed servitude, enclosures, aggregating ownership to one privileged class and clustering waste and pollutants near the places of the disempowered. Propriety, however, demands the good treatment of both people and earth. Liberty Hyde Bailey, an early twentieth-century American horticulturalist, put it this way: "One does not act rightly toward one's fellows if one does not know how to act rightly toward the earth."[81]

And by acting rightly toward both, we create the foundation for what we all need—a healthy economy. Economists call this the triple bottom line. (It is alternately referred to as People, Prosperity, Planet, or Earth, Economy, Equity.) A healthy environment depends on and gives rise to healthy workers, who in turn depend on and give rise to a healthy economy. It's a win-win-win. Abuse or deplete the earth and its resources, and both the economy and the people suffer. All depends on how we treat the earth. As the world-class economist Herman Daly put it, the economy is a subset of the natural environment, not the other way around. "The

Enclosure is the practice of closing off common land on which villagers had centuries-old rights to graze their cattle, hunt, and forage for food and fuel, depriving them of this age-old right. Being removed from sources of food and energy often forced self-sufficient families to become laborers, dependent on harsh overlords, or pay exorbitant fees for what used to be shared gifts of nature. The most infamous enclosure movement happened in Britain in the eighteenth and nineteenth centuries, though it was practiced elsewhere as well. Enclosure was usually enforced with walls, fences, or hedges and often violence.

value of a sawmill is zero without forests; the value of fishing boats is zero without fish; the value of refineries is zero without remaining deposits of petroleum; the value of dams is zero without rivers."[82]

Seeing the Sparks

When we get caught up in the trends and treadmill of the everyday, we often lose sight of the sparks that are all around us. Yet remembering them can re-center and soothe us, moving us to care for the places where the sparks reside.

Many blessings in Judaism enable us to do just that, to slow down, look, and attend to the wonders of the world. Blessings are rituals that urge us to pause, see, and appreciate the sparks. They are nudges, tugs, reminders not to let the possibility of awe pass unheralded. In Heschel's words, they are meeting places between God and self.

In the month of Nisan, the month of Passover and the advent of spring, we are to say a prayer upon seeing the first blossoms of fruit trees: "Blessed are You, God, Creator of all, who has made nothing lacking in this world and created in it goodly creations and goodly trees so we may gain pleasure from them." When we see the ocean for the first time ever or after a long while, we say, "Blessed are You, God, who made this great sea." There are blessings to be said over thunder, lightning, comets, and unusual phenomena. There are blessings to be said upon eating new foods and upon seeing the gathering of thousands of different people, each unique in their own way. There are blessings that announce the new moon and celebrate the waxing moon; blessings to be said over light and aromatic spices. There is even a blessing that is recited once every twenty-eight years,

Blessings of Awe

Upon seeing lightning, majestic mountain ranges, and other remarkable natural phenomena, we say,

"Blessed are You, Adonai our God, Source of all, who continually performs acts of creation."

Upon hearing thunder, we say,

"Blessed are You, Adonai our God, Source of all, whose power and might fill the world."

Upon sensing a pleasant scent, we say,

"Blessed are You, Adonai our God, Source of all, who creates all sorts of wondrous fragrances."

Upon seeing the waxing moon, we go outside and say,

"Praised are You, Adonai our God, Source of all, whose word created the heavens and all its hosts with a single breath. You assigned them their appointed times and tasks. They did not alter their mission, doing their Creator's bidding with gladness and joy. You are the Creator of truth, whose creation is truth, and have said to the moon that it shall always be renewed. It is a crown of splendor for those borne by God, for they too are destined to be renewed like the moon and proclaim the beauty and glorious majesty of their Creator. Praised are You, Adonai, who renews new moons."

Awe unites body and spirit, self and earth, even soul to soul. It allows us to know we are more than our bounded selves; that we belong to something grander, enduring, transcendent; that though we are small, we are part of something very large. Awe can introduce us to infinite secrets.[83]

when the sun—according to rabbinic calculations—returns to the exact position it assumed at the moment of its creation.[84] As with all these, we are called to notice.

Rabbi Eugene Borowitz, a teacher and philosopher of the late twentieth century, who seemed to always have a sparkle in his eyes, urged us to ask ourselves "with full force and fervor: Why is there anything at all?"[85] That there is life all around us in great abundance, with explosions of color, smell, size, and shape, some glorious, some not, should not inure us to its improbability or numb us to the surprise of it all. It should not fail to astonish us. Indeed, as the myriad blessings we are to say throughout the days, weeks, months, years remind us, it is a remarkable thing that we, life, any of this, exists.

Whether through serendipity or design, how amazing is it that we have liquid water in the right amounts and the distributions that we do; that our planet is located in the solar system where it is (the Goldilocks zone); that we have an atmosphere that protects us from the destructive rays of the sun and debris from space; that we have the right amount of oxygen and the lungs to fill with it; that we have plants and fungi and bacteria that grow and renew all that dies; that we have the moon and the tides; that we are on a hospitable arm of our galaxy—all this and more all at the same time.

Abraham ibn Ezra, the twelfth-century poet, linguist, translator, and Torah commentator from Spain, saw nature as a place to encounter God. He wrote a poem, a blessing of a sort, celebrating such meetings:

> Wherever I turn my eyes, here on Earth or to the heavens
>
> I see You in the field of stars

I see You in the yield of the land
in every breath and sound, a blade of grass, a simple flower, an echo of Your holy Name.[86]

In the grandest of the grand and the smallest of the small, Ibn Ezra saw God. Not God's very self (God and nature were not to be conflated), not even God's name, but an echo, an imprint, a whiff, a telltale sign of the universe's Maker. And that made his world magical.

It is, in part, the mystery of life, its unfathomable quality, that makes it so intoxicating. Like so many who have been captivated by this world, G. K. Chesterton, a British essayist and social commentator, did not demand an explanation as to why or how. He just took it all in: "I felt in my bones; first, that this world does not explain itself. It may be a miracle with a supernatural explanation; it may be a conjuring trick, with a natural explanation. But the explanation of the conjuring trick, if it is to satisfy me, will have to be better than the natural explanations I have heard. The [world] is magic, true or false."[87] No matter the reason, life is astonishing. The deeper we penetrate life's mysteries, the more astonishing it gets. Yet it is not necessary to understand. To witness is enough.

Neither logic, nor science, nor piety can (yet) fully explain it, and we should not contort them to do so. Our task is to accept that we are living in the midst of a grand and unknowable wonder. As Leonard Cohen explained about why he wrote his massive[88] and massively popular song "Hallelujah": "You look around and you see a world that cannot be made sense of, you either raise your fist or you say hallelujah."

Perhaps that is the odd comfort the end of the Book of Job is meant to evoke. Throughout the book, Job

Poet Leah Goldberg asked for help seeing and feeling the sacred, for while knowing awe may come naturally to some, others can use a little assistance:

Teach me, my God,
to bless
and to pray
Over the secret of the
withered
leaf, on the glow
of ripe fruit,
Over this freedom: to see,
to feel, to breathe,
To know, to wish, to fail.
Teach my lips blessing
and song of praise,
Renewing your time each
morning, each night,
Lest my day today be
as days gone by
Lest my day become
for me
simply habit.[89]

Perhaps too the Book of Job offers an answer in this way: In the end, God shows up. God responds. Knowing there is a God whose ways are inscrutable to humankind seems like it might be the best comfort the author can muster.

rejects his friends' suggestions that he is responsible for the tragedies he is suffering, denying that he did anything to deserve them. He challenges God, demanding to know why this is happening. At the very end of this long book, after many chapters of Job's friends' futile arguments, God appears. God summarily dismisses Job's friends' arguments and confronts Job with a seemingly endless avalanche of questions: "Where were you when I laid the foundations of the world… when I said [to the seas], 'You may come so far and no further…'? Have you ever commanded the day to break, assigned the dawn its place?…Can you tie cords to the Pleiades or undo the reins of Orion?…Can you send an order up to the clouds…or dispatch lightning on a mission?… Does the eagle soar high at your command?" (Job 38:4, 38:11–12, 38:31, 38:34–35, 39:27). The questions continue for many verses.

We cannot understand the ways of the world, the author seems to be telling us. "I spoke without understanding" (Job 42:3). But for whatever period of time we tread this earth, we are participants in this immense, engaging, enraging mystery of Life, swirling all about us. And that, the Book of Job seems to be telling us, must be reward—and answer—enough.

We may not understand why we are here or how we got here, but here we are. And we can do what no other creature on earth can do. We can serve as witness. No other creature, as far as we know, can stand back and take in the wonders of the universe. No other creature can be awed as we are by the changing of the seasons, the song of the nightingale, the vastness of the heavens. No other creature can contemplate ancient beginnings or comprehend the concept of eternity or tell stories about how we came to be or why. To be gifted with this ability, to dare to peer into the heart of the universe, to feel the joy of wonder despite the pain of living and the fear of dying is to know amazement indeed. As Carl Sagan said, "We are a way for the universe to know itself."[90] It is this—our appreciation and reverence for the world's wonder and enchantment—that can best lead to its preservation.

A Torah of Nature

The sense that Creation was the beloved work of God and that we could therefore discover the Divine in all aspects of nature erupted in an enigmatic little book known as *Perek Shirah*, meaning "a litany of verse." Sometime over the long years of exile (we don't know when or where), an anonymous author was moved to weave together the lessons and love of nature with the lessons and love of Torah.[91] What emerged is a short, pithy catalog of biblical verses mentioning various elements of nature—heaven and earth, mountains and sea, animals and vegetation—which reveal how nature in all its diversity celebrates God.

The message of *Perek Shirah* is that just as all of nature praises God, so should we. The world is not silent; it professes. Nature is not dumb. It is full of song, which we too can hear if we but listen. *Perek Shirah* takes the reader on a Jewish tour of the universe, pointing to the sacred within and beyond. "The heavens are saying: 'The heavens tell of the glory of God, the firmament of God's handiwork' [Psalm 19:2]."[92] "The vegetables of the field are saying: 'You water its furrows abundantly, making its ridges soft with showers, blessing its growth' [Psalm 65:11]."[93] Throughout the more than eighty entries, nature appears before us in a riotous array of color and style and matter. The simple structure of the text—one line for each element—allows for its quick review, leading to a tradition that promises if you say *Perek Shirah* for forty days as a spiritual discipline, you will see miracles. Even now, some people recite *Perek Shirah* daily.

In the sixteenth century, many years after *Perek Shirah* entered our liturgical canon, on a hilltop in Israel, the denizens of Safed, led by Rabbi Isaac Luria, infused nature with even greater meaning. These were the kabbalists, mystics who believed nature did not just make us aware of the Divine, or sing praises to the Divine, or give witness to the Divine. Even more, nature was "ensouled" with the Divine.[95] This, they explained, is how it happened: Once upon a time, the Ein Sof, the One who is all, who has neither beginning nor end, who is beyond time, dimension, and all description, chose to create this material world. But there was a problem. The Ein Sof filled the universe, leaving no room for the matter of Creation. So the Ein Sof did the unthinkable. It contracted, withdrew a bit of the infinite self, fashioning a space to bring the universe into being.

Into this emptiness, vessels of matter were crafted and arrayed, forming a sequence of descending channels through which the animating light would flow from the Ein Sof into the physical world. When all was set, when the vessels were in place, the light of the Divine began to flow through this structure. But despite the best planning by the Architect on high, the celestial brilliance overwhelmed the vessels, shattering them and scattering them, forming the imperfect world we know today.

Still, the sacred was not lost. Within each material shard rests a spark of the divine light, embedding sacredness throughout this broken, physical world. Our

Torah and the Text of Nature

This notion of nature being a worldly torah animating the written Torah is reminiscent of a passage penned by the great American nature writer John Burroughs. He was almost certainly unfamiliar with the writing and typography of classic Jewish texts, yet he penned a paean to nature that precisely conjured up images of such Jewish texts, further blending the torah of nature with the Torah of text. "The book of nature," Burroughs explains, "is like a page written over or printed upon with different-sized characters and in many languages, interlined and cross-lined, and with a great variety of marginal notes and references. There is coarse print and fine print.... We all read the large type more or less appreciatively, but only the students and lovers of nature read the fine lines and the footnotes. It is a book which he reads best who goes the most slowly or even tarries long by the way... [and] dwells fondly upon its most obscure text."[94]

This could be a description of classic Jewish texts: multilayered with different fonts; a jumble of generations captured on the same page in constant conversation, seeking truth. Many students of Jewish texts thoroughly immerse themselves in the words, tarrying among them, savoring the adventure, dwelling on the obscure, ferreting out the most subtle of meanings, reveling in the finest of nuances. Jewish devotees of nature, like the author of *Perek Shirah*, the psalmists, the poets, the mystics, do the same—seeking, reading, and reveling in the torah of creation, teaching us about life, God, and ourselves.

task, we are told, is to liberate those sparks, to set them free, allowing them to return to their source and thus repair the primordial rupture of the world. We are to do that by mindfully using the matter of this world to fulfill the observance of the commandments, in the proper way, accompanied by the proper intention. For the material world is not to be shunned or denied or denigrated or even transcended, but rather to be encountered in a sacred way. Though broken, it is still a conduit of the Divine. It is up to the spiritually attuned among us to use it well, incorporating it into our rituals, thereby effecting the repair of the world (*tikkun olam*).

In concert with the kabbalists' belief in the sympathetic engagement between earth and the Divine, the first Tu BiShevat seder Haggadah was crafted, called *P'ri Eitz Hadar*, meaning "the goodly fruit."[96] First published in Venice in 1728, it was modeled on the Passover Haggadah, with rituals and a guiding narrative that took the celebrant through a sacred, mystical meal. In the days of the Temple, Tu BiShevat had been an accounting tool denoting the new year of the trees, that is, the boundary between one arboreal tax year and the next. Over the Jewish people's many years of exile, it had been largely ignored, preserved as a mere historical artifact with no contemporary application. The kabbalists transformed this minor day of recordkeeping into a joyous theurgic celebration, supercharging the flow of holiness between heaven and earth.

The text of this seder begins:

> Please, God, You who make, form, create, and emanate supernal worlds, who created their likeness on the earth below, according to their supernal forms above and their lower forms be-

> low, join these two worlds together in a single shelter so all will be one. You have caused trees and grass to grow upon the earth, according to the structure and character of the forms above, so that human beings might gain wisdom and understanding through them, and thus grasp what was hidden. You appointed Your holy angels over them as agents to oversee their growth. And You caused *shefa* [the supernal flow] and the power of Your supernal qualities to rain down upon them.

And on it goes. The seder recognizes the natural world as an earthly avatar of the eternal world. The fate of the two are intertwined, the lower world determining the fate, as it were, of the upper world. By eating various fruits with different structures (inedible outsides; inedible cores; all edible), by invoking the right *kavanah*, prayer of intention, the primordial rupture may be repaired and the reunification of the lower world and the upper world hastened.

The legacy of this tradition lives on today. Countless Tu BiShevat Haggadot have been written over the past few decades; countless more Tu BiShevat seders have been celebrated all over the world. And the phenomenon seems to be growing. While the spiritual, even mystical, elements of these modern creations are often retained, they are increasingly being joined by urges toward environmental advocacy. Today's celebrants are not just celebrating the ways of nature but promoting actions to protect it. They are connecting their love of Judaism with their concern for the state of the world. And they are doing this through tradition, community, ritual, and diverse food delights. The seder is a way for the participants to reclaim and renew the covenant between earth and heaven and between earth and people. Through the seder, they hallow and inspire acts that care for the world.

One hundred years after *P'ri Eitz Hadar*, no less a luminary than Rabbi Nachman of Bratslav, the eighteenth-century mystic, teacher, and great-grandson of the Baal Shem Tov, sought closeness to the Divine in nature. Like his great-grandfather before him, Rabbi Nachman was an honorary denizen of the wilds. He went there to lose the constraints of self-consciousness that society often imposes and to open himself fully and trustingly to God—and to recruit nature to help complete his task.

One of the most famous prayers ascribed to Rabbi Nachman is the following:

> May it be my custom to go outdoors each day among the trees and grass—among all growing things and there may I be alone, and enter into prayer, to talk with the One to whom I belong. May I express there everything in my heart, and may all the greenery of the field—all grasses, trees, and plants—awake at my coming, to send the powers of their life into the words of my prayer so that my prayer and speech are made whole through the life and spirit of all growing things, which are made as one by their transcendent Source."[97]

It is as if both Rabbi Nachman and the earth come alive in one another's presence. The prayer of the supplicant animates nature as nature elevates and animates the prayer of the supplicant. The one gives wing to the other and offers a refuge from human-induced constraints. The safety, the solitude, the gracious welcome we find in nature's gentle wildness awakens us and rouses in us the powers inherent in growing things. The world, he counsels, is kissed by God. How can we do otherwise than care, tenderly, for such a world?

Rabbi Abraham Isaac Kook, the popular, mystical, politically astute, twentieth-century rabbi of pre-state Israel, was a spiritual disciple of these sentiments. He too believed nature not only to be inspirited by its transcendent Source but able to impart that spirit to us. Nature rings with a divine symphony, he taught, stirring and inspiring us, if only we will allow ourselves to listen. "Each grass says something, each stone whispers a secret, and each being sings a song."[98] The sacred is thus not content simply to be. It yearns to be known, to share its truth. And Judaism is a willing partner to this knowing. "The Jewish vision," Rabbi Kook writes, "is the holiness of all creation."[99]

Israel's Holy Sparks

Over the years, the Jewish people's celebration of nature has continually been stitched into the fabric of our sacred texts, our prayers, our commentaries, the mystical traditions that punctuated our imagination, the words and poetry of many of our most beloved teachers. Although, as was noted, this connection to nature was often constrained by some schools of Judaism, it again and again found a way to burst through and enchant the spirit of the Jewish people, which it did mightily in the early modern Zionist movement. The *chalutzim*, the pioneers who led the way back to the Land of Israel in the late nineteenth and early twentieth centuries, believed the land had a life-giving force. In the

One of the greatest *chalutzim* (pioneers who emigrated to Israel in the early twentieth century) to champion the sacredness of working the Promised Land and the transcendent feelings born from it, was A. D. Gordon. "You work here simply," he wrote, "without philosophizing; sometimes the work is hard and crowded with pettiness. But at times you feel a surge of cosmic exaltation, like the clear light of the heavens. Unfathomed depths stir within you. And you, too, seem to be taking root in the soil which you are digging, to be nourished by the rays of the sun, to share life with the tiniest blade of grass, with each flower; living in nature's depths, you seem then to rise and grow into the vast expanse of the universe."[100] Eternity, purpose, peace could be found turning the rich soil of the Galilee.

words of an early Zionist song, "We came to build up this land and to be built up by it." From the land and their work on it, the *chalutzim* believed they were reaping redemption for themselves and the millions of Jews who suffered through the long, harsh centuries of exile.

Distinct from their ancestors, though, they were not so much intoxicated by the godliness of the land as by its earthiness—the soil, the air, the smells. It was a land they could tend to, which in turn would tend to them; a land they could call home and from which no one could displace them.

I was in Hebrew school in the 1960s, those days when the realization of the Zionist dream was still fresh and new. One year, the students in our class were assigned pen pals from Israel. We wrote to them about our lives in America, and they wrote to us about their lives in Israel. I remember nothing of what we said, but I can never forget what accompanied each letter: a delicate wildflower that my pen pal had picked and pressed. Every classmate got a flower in their letter too. We never thought of sending oak leaves or pressed echinacea or black-eyed Susans to our pen pals in Israel. But in every letter we received from them was a gift that heralded the beauty of the Land of Israel and our pen pals' pride in it. (Wildflower picking was so popular in Israel that it had to be banned, giving new meaning to love them and leave them.)

Rahel Bluwstein was a poet who wrote of the early Zionists' captivation with the land. Like many of the *chalutzim*, Rahel was born in eastern Europe. She made *aliyah* (moved to Israel) in 1909, when she was only nineteen. Settling at first in Kibbutz Kinneret, in sight of the Sea of Galilee (the Kinneret), she devoted herself to learning the ways of the land. But, after World War

I, she contracted tuberculosis, and her lingering poor health forced her to leave her work and the Galilee she loved. Distanced from her beloved landscape, she wrote a heartrending poem speaking to the land and lake that had given her life but which she would see no more. The poem has been set to music and continues to be sung to this day.

> Perhaps it was never so.
> Perhaps
> I never woke early and went to the fields
> To labor in the sweat of my brow
> Nor in the long blazing days
> Of harvest
> On top of the wagon laden with sheaves,
> Made my voice ring with song
> Nor bathed myself clean in the calm
> Blue water
> Of my Kinneret. O, my Kinneret,
> Were you there or did I only dream?[101]

Images of nature have saturated the modern State of Israel. The seven-branched menorah on Israel's official emblem is said to be modeled on the seven-branched, aromatic native plant *Salvia palaestina*. The olive branches (olives being native to Israel) surrounding the menorah on Israel's state seal symbolize the hope for peace.

From Text to Action

Thousands of years of such enchantment with nature continues in an explosion of Jewish environmental activists today. Inspired by the psalms, prayers, Hebrew poetry, and the holidays, environmental activ-

Saul Tchernikovsky was another celebrated early Zionist poet who believed nature offered a sacred encounter with the essence of all being. This was not a mystical experience available only to the chosen few. The divine exists, he professed, in the here and now. We walk in its shade, drink its waters. The divine is in all we do, in the everyday cycles of birth and death and birth again; in the ways wadis suddenly fill with water, like flashes of inspiration, roaring with rains that fell miles away.

> And if you ask me of God,
> my God
> 'Where is God
> that in joy we may
> worship?,'
> Here on Earth too
> God lives,
> not in heaven alone....
> Wherever the breath
> of life flows, you will
> find God embodied.
> And God's household?
> All being:
> the gazelle, the turtle,
> the shrub, the cloud
> pregnant with thunder...
> God-in-Creation
> is God's eternal name.[102]

Just the other day, my four-year-old grandson was singing a tune I did not recognize. It was some simple expression about the glory of being, as many children's songs are in their own way. When I asked him what he was singing, he just shrugged and said, "I have a lot of songs in my life." May we all be blessed with lots of songs.

ists are translating the spiritual and mystical into social and political action. Throughout the Diaspora as well as in Israel, Jewish farmers are pursuing regenerative agriculture that allows the land to become richer with every harvest, increasing its yield so that none go hungry. They are working to mitigate the impact of climate change by reducing emissions of greenhouse gases and enabling the soil to absorb more carbon dioxide.[103] Many others are turning away from the chimera of a world that confuses growth with progress. They imagine a dynamic world in which we consume less, recycle more, all the while building a more just, fulfilling, sustainable market. They seek to purge society of our legacy of waste, knowing that all of nature is made of cycles, where the end of one life serves as the beginning of another. They know this can happen if we all reclaim a Genesis 2 relationship with the earth, so we may once again see the holy sparks in a blade of grass. In that way, we can achieve the task of *yishuv ha'olam* that we are all called to do, as the sixteenth-century kabbalist Rabbi Moshe ben Machir wrote: to "leave a blessing after ourselves, so that 'generation after generation, the world remains forever.'"[104]

CHAPTER FIVE

The Limits to Private Ownership

We love our things. We love owning them, buying them, getting them, holding them; we love wearing them, collecting them, organizing them, displaying them, polishing them, adorning them, describing them. We love how our things make us look and how they make us feel. Perhaps most of all, we love knowing they are ours alone and that we are able to do whatever we want with them. From the moment we clutch our toys in our little hands and desperately yell, "Mine," while our parents, just as desperately, urge us to share, we express the desire to exercise our right of "despotic dominion" over things.[105]

And no wonder. Things do so much for us. They tell a tale about us, speaking for us even before we utter a word. That is why doctors wear white coats and judges black robes. It is why teenagers all dress the same and men wear power ties. It is why some Jews wear head coverings, and why there are so many types (the knitted *kippah*, the black hat, the furry round *shtreimel*, the scarf, the wig, and more). It is why we carry the handbags we do and the water bottles we use. It is often why we choose the shoes we wear, the cars we drive, even the foods we eat and the coffee we drink. They are touchstones, visuals of the stories we seek to tell about ourselves. We use them to bind body and soul. They offer up the inner parts of self we seek to share with the outer world while sheltering the parts that are most vulnerable. In short, wielded wisely, our things reveal and conceal in well-calibrated ways and amounts.

Even more, in this mercurial and unpredictable world, our things form an island of security and refuge. As Hannah Arendt, a twentieth-century political and social commentator, wrote, "The things of the world have the function of stabilizing human life.... [People], their ever-changing nature notwithstanding, can retrieve their sameness, that is, their identity, by being related to the same chair and the same table."[106] That is to say, being surrounded by our things soothes us like a meditative practice. Our things center us by the comfort of their constant presence, reassuring us of who and what we are, no matter what the world threw at us that day.

Judaism recognizes this bond between spirit and matter. It calls us to wrap ourselves in a tallit, a large four-cornered prayer shawl with fringes dangling from each end, so we may daily be reminded of who we are and what we are to do. It calls us to adorn the doorposts of our homes with *mezuzot*, a decorative case containing words of Torah, so we may be guided in our comings and goings by the teachings of our tradition. It calls us to dwell in *sukkot*, makeshift huts, for one week of the year, to remind us of life's fragility and nature's gifts.

Judaism honors and protects our right to material possessions. The eighth commandment forbids stealing; the tenth forbids even coveting the things of other people (Exodus 20:13–14). We may not move another's boundary marker (Deuteronomy 19:14; Proverbs 22:28); we may not keep a poor person's garment (which we have as collateral for a loan) overnight but must return it at day's end, for it may be the only garment that the person has to sleep in (Exodus 22:25–26; Deuteronomy 24:12–13). Embellishing on all this, the ancient rabbis devoted an entire book of the Talmud to examining the complex world of property law and the protocols of returning lost objects.[107]

The Earth Is God's

And yet, despite this acknowledgment of the rights and value of our material objects, Judaism limits the authority of private ownership, countering much of the Western world's ideas about our rights over our things. To better understand why and how, it is helpful to distinguish between three categories of property: ownership, possession, and user rights.

Ownership, in Western society, historically gives the owner despotic rights—to the exclusion of anyone else—to determine the use of the object owned.[108] The owner, for most intents and purposes, is assumed to have complete jurisdiction over what they own. They may deny anyone else access to or benefit from it. They may employ and dispose of the object as they desire. They may use it wisely or wantonly; they may save it or destroy it. If they decide to burn their dining room chairs to make s'mores, or cut their wedding dress into rags to wash the kitchen floor, or destroy anything they own in a pique of anger, a display of wealth or power, or for no reason at all, that, according to Western legal traditions, is totally within their right.

Possession, however, is different from ownership. Possession is about being in the presence of and having physical control over and immediate access to something, yet without the full rights of ownership. Not everything we possess do we own, and not everything we own do we possess. Landlords own rental units, but renters possess them. Many museums exhibit paintings that they possess but do not own. When we borrow books from the library, we possess them, but we don't own them. Neither renters nor museums nor book borrowers may alter or dispose of the items they possess as if they were their owners. Possession is constrained by the terms of use or loan, while ownership is largely unconditional and unconstrained.

And then there are user rights.[109] Offering neither the privileges of ownership nor full rights of possession (though possession often includes user rights), user rights confer the ability to use and benefit from an object or land. Drivers, for example, have the right to use street parking and the streets themselves (sometimes for free, sometimes for a fee), though they neither own nor possess that space. Pedestrians have the right to walk on a city sidewalk, though they too neither own nor possess the sidewalk.[110] Farmers who lease land pay for the right to farm it, but they neither own nor possess it. Buying a cup of coffee in a café affords patrons the right to sit at a table, though they neither own nor possess the store, the mug, or the furniture they sit on.

The rules of ownership, possession, and user rights are distinct, but the boundaries between them differ across cultures. The "right to roam," for example, may give a person in some countries limited access (by way of user rights) to what is otherwise private property, including using it as a shortcut, as a pathway to a water source, or access to forage the plants that grow on it. Many Nordic and European countries allow such rights today. The Torah too, speaks of this right, teaching, for example, "When you enter a fellow's vineyard, you may eat as many grapes as you want, until you are full, but you must not put any in your vessel" (Deuteronomy 23:25). It is fine to enter a neighbor's property and grab a bite to eat when using his field as a shortcut, but there is to be no harvesting, no carrying anything away, no harming, and no dawdling. In other cultures, however, such acts might be considered trespassing and theft.

The Torah knows of these three ways of dividing up and allotting the rights to things, but it presents a different theory of ownership: "The earth is the Eternal's and the fullness thereof" (Psalm 24:1). That is the essence of all material being and the starting point of all material use. This is not simply a pious pontification for the psalmist. It is a statement that frames the tradition's approach to humanity's relationship with the physical world: everything belongs to God. All things emerge from God's act of Creation. "All is from You," I Chronicles (29:14) has King David say regarding the Temple. But it pertains to all matter. "All that we have laid aside to build You a House for Your holy name is from You, and it is all Yours" (I Chronicles 29:16).

All things belong to the Creator. Ownership resides not in us but in God. Therefore, despotic ownership can never be ours. Possession and right-to-use are all we can claim. Tradition gives us the right to hold, manipulate, alter, refine, and seek to improve things, but we don't and can't fully own things.

For we don't and can't create matter. Everything we touch preexists us. All we can do is use what we find here, all of which can ultimately be traced back to its source. The fabrications of human ingenuity (both material and spiritual) all rely on what we find here; all are born of the preexisting matter of the world. Anthropologist and educator Loren Eiseley wrote, "The human mind, so frail, so perishable, so full of inexhaustible dreams and hunger, burns by the power of a leaf."[111] The thoughts we conjure up, the laptop I am typing on, the energy you are putting in to read this sentence, the design of our cities, our clothes, our pens, our toys, our mailboxes, each are products of the bounty of the earth.

Imagine, for example, a farm. We may work it, fence it, plow it, seed it, weed it, sweat over it. And at the end of the day, we may claim the final product as ours. But even as we do, we cannot forget that we did not make

For believers, the source of all Creation is God. For others, the world may have come into existence through happenstance, serendipity, or some process equally mysterious. Either way, the teachings of humility in relation to the natural world holds. We do not create, only manipulate. When we overstep our bounds, we put ourselves and the world in peril.

the stuff that does the work. We did not make the soil, the rain, the sun, the climate that is so conducive to growing this crop. We did not make the matter that our farm implements are made of. Having this awareness and letting it guide our engagement with earth's goods is what Jewish tradition urges us to remember.

It is true that the psalms also say, "The heavens belong to the Eternal, but God gave the earth over to humankind" (Psalm 115:16). Couldn't we rightfully assert, then, that this gives us full rights over the earth? Doesn't that create a division of realms, wherein God holds sway over the heavens but cedes the earth to us, justifying human hegemony over global resources? Ibn Ezra, the eleventh-century Spanish biblical commentator, anticipated such a challenge and responded this way: "One might think this means that just as God rules the heavens so humans can rule the earth. But that is not so. God's kingdom encompasses all. Rather, what the verse means when it says, 'God gave the earth to humankind,' is this: that humanity is like a caretaker for God on earth and all it possesses. For all is according to the word of God."[112] God never relinquishes ownership over the earth, Ibn Ezra and others argue. Rather, God appoints humans to stand in God's stead, as God's agent, to care for and protect the well-being of the earth.

In return, we are given the right to use it wisely, *l'ovdah ul'shomrah*, to till it and to tend it, to work and protect it, per Genesis 2. We can never take this right or the earth for granted. As a constant reminder of this partnership, the Jews of the biblical period were bidden to offer back to the Creator a portion of earth's gifts that they worked hard to bring forth. Throughout the time the Temple stood, Jews brought their animals, grains, and fruits—those things that grew under their watch—as sacrifices to God, grateful for the abundance that God and the earth had given them. The pilgrimage holidays (the days on which the Israelites made pilgrimage to Jerusalem to offer their holiday sacrifices, those being on Passover, Shavuot, and Sukkot) revolved around the seasonal harvests. Year after year, the first fruits and firstborn animals were dedicated to God, giving the household the right to benefit from the rest of the harvest and flock and reminding the homesteader to whom the land truly belonged.

Jewish practices reflect this attitude today as well. We offer blessings before and after eating, acknowledging our gratefulness to God, to the fecundity of the earth, and, we might add, to all who participate in the long supply chain connecting the source of food to our table. Not reciting such blessings, the Talmud tells us in Berakhot 35a, would be as if we were stealing from God and the earth. And though such an attitude is codified in the liturgy around eating, it is designed to guide all that we have. In an extreme but nonetheless heartfelt act of piety, some observant Jews have bookplates that riff off of Psalm 24:1. It acknowledges that all things have their source in God's Creation and that God therefore is the rightful owner of all. Yet it also recognizes the necessity of possession and that user rights are given to us: "The earth is the Eternal's and the fullness thereof," it says, "but this book for now resides in the library of __________."

This concept that only God is able to claim true ownership of matter on earth is fundamental to the Jewish theory of use. It plays itself out even in areas that contradict the most sacred rules of Western thought. John Locke, who had a profound influence on how the

West thinks about private property, argued that once a person added their own labor to the elements they wrested from the earth, that person, by rights, possessed exclusive ownership of what they made.[113] American law tends to follow this belief, protecting us from being deprived of the full value, potential, and fair use of what we own. Limiting our enjoyment of our property ("takings" in law) can only be justified through due process, wherein the government determines that such appropriation of property, or limitation of property use, is necessary for the betterment of all and assures that the owner will be duly compensated for any lost value.

Judaism teaches differently.

The Imperative of "Wise Use"

The belief that we do not enjoy absolute rights over what we possess resounds throughout the Bible and rabbinic law. The Torah's most famous articulation of this teaching—as understood and expanded upon by the rabbis—is that of *bal tashchit*, "do not destroy." "When you are making war against a city, and you place it under siege for a long time in order to capture it, you must not destroy its [fruit] trees, wielding the ax against them. You may eat of them, but you must not cut them down. Are trees of the field human to withdraw before you into the besieged city? Only trees that you know do not yield food[114] may be destroyed" (Deuteronomy 20:19–20).

These verses appear in the midst of a text that deals with various laws of warfare. All war is regrettable, yet not all war is avoidable. The challenge is how to pursue the undesirable in the most judicious way possible. The Torah tells us that war does not give us permission to pursue a scorched-earth policy. As one reading of these

Locke's Caveats

And yet, it seems the Western world has ignored two of Locke's caveats that constrain such absolute rights.

The first: For those things that we have made or improved, Locke writes, we enjoy exclusive rights, "at least where there is enough, and as good, left in common for others."[115] That is, even those who wish to abide by Locke's rule of ownership gained through labor should agree that the right of such ownership is constrained by the call for the equitable access and distribution of the earth's resources. One's industriousness and the ability to be there first do not allow one to call "dibs" to the earth's goodness if one thereby prevents others from enjoying equal access to that goodness.

The second: Locke warns against environmental injustice: "But though this be a state of liberty, yet it is not a state of licence: though man in that state have an uncontrollable liberty to dispose of his person or possessions. . . . The state of nature has a law of nature to govern it, which obliges every one:... no one ought to harm another in his life, health, liberty, or possessions."[116] That is to say, common sense teaches us that my freedom to use my resources as I wish stops at the point where my use harms you. My choice to drill, mine, frack, burn things, destroy things, manufacture things, build things, even grow things on my property in a way that ends up harming you is not permitted. There is a movement in the United States and around the world to enshrine such environmental human rights in the fundamental laws of the land.

verses suggests (for the Hebrew lends itself to some ambiguity), the fruit tree is a defenseless noncombatant, innocent of any threat. It does not deserve to be a pawn in this war. (In the translation provided, the reason given to preserve the tree is that it cannot defend itself by retreating into the city. Both readings render the same conclusion: fruit trees have "standing" and must be protected, even in the midst of war.) Even more, permanently destroying life-giving elements of the earth—upon which our future well-being relies—for temporary gains is a bad calculus and forbidden. Tomorrow cannot be sacrificed on the altar of today.[117]

And if this is true in war, how much more so in the rest of life? If the demands of combat do not justify violating and destroying earth's goodness, surely the demands of ordinary life don't either. Over the centuries, this singular prohibition of unwise and unethical destruction in the midst of a specific condition of warfare has expanded to become the foundation on which much of the edifice of Jewish sustainability has been built. Rabbi Samson Raphael Hirsch, the influential nineteenth-century German scholar and Judaism's greatest and most passionate proponent and explicator of *bal tashchit*, the prohibition of wastefulness, wrote in his classic commentary on these verses in the Torah:

> This prohibition [of destroying fruit trees in a siege]...is to be taken as an example of general wastefulness. Under the concept *bal tashchit*, the purposeless destruction of anything at all is taken to be forbidden, so that...our text becomes the most comprehensive warning to human beings not to misuse the position which God has given them as masters of the world and its mat-

> ter to capricious, passionate or merely thoughtless wasteful destruction.... Only for wise use has God laid the world at our feet when he said to Man "subdue the world and have dominion over it."[118]

To most of us, waste is a natural by-product of life. It seems inevitable, pervasive, carrying no negative moral or ecological implication. It is not obviously paired with the concept of destruction and is excused as an accidental, even unavoidable, by-product of human activity. But it need not be and should not be. "Waste," after all, "is not present in nature. Waste is the product of a linear system: Materials are extracted from the earth, processed and used, often very briefly, before being thrown out. Ecosystems, though, have circular systems: Materials are used and recycled in constant loops."[119] Waste, in short, is a design failure, a failure of human imagination.

To Hirsch, waste was more than a flaw; it was a sin. In that mindset, waste was seen not simply as unusable excess that is readily, unceremoniously, and often thoughtlessly discarded, but as "an act or instance of using or expending something carelessly, extravagantly, or to no purpose."[120] Increasingly those engaged in promoting sustainability see waste this way too, as unnatural and destructive, a problem to be solved.

Hirsch does not exhaust all he has to say about waste and destruction in his Torah commentary. In *Horeb*, his two-volume magnum opus on Jewish practice, he devotes three full pages to *bal tashchit* and the ethics that undergird it. For Hirsch, *bal tashchit* is the premier mitzvah upon which all the rest of God's commands depend:

Hirsch sees *bal tashchit* as *the* foundational mitzvah. *Sefer Hachinuch* sees *yishuv ha'olam* as *the* foundational mitzvah. There is no contradiction. They both point in the same direction and meld into one. *Bal tashchit* begins with a particular negative command ("thou shall not destroy fruit trees...") and expands into a categorical teaching of no waste of any kind. *Yishuv ha'olam* begins with a categorical positive command ("thou shall...") which we are to translate into particular behaviors that yield a healthy world. They are two sides of the same coin.

> "Do not destroy anything!" is the first and most general call of God.... All around, you perceive earth and plant and animal... already bearing your imprint from your technical human skill.... They have been transformed by your human hand for your human purposes.... If you should now raise your hand... wishing to destroy that which you should only use... if you should regard the beings beneath you as being objects without rights, not perceiving God Who created them... then God's call proclaims to you, "Do not destroy anything!"[121]

Hirsch acknowledges that humans have great power and, by right, are able to use that power, but only to the extent conferred by wisdom and justice and the belief that all belongs to God, put here for the needs of both those alive today and the generations yet to come. "Use the things around you for wise purposes," Hirsch imagines God telling us. "I lent them to you for wise use only; never forget that I lent them to you."

Hirsch echoes the rabbis of the Talmud who first extended the rules of *bal tashchit* beyond protecting fruit trees in war, even beyond elements of nature to include human-created items, whether those be clothes,[122] utensils, books, furniture, anything. "Under the concept of *bal tashchit*, the purposeless [reckless, excessive, or unnecessary] destruction of anything at all is taken to be forbidden so that the *bal tashchit* of our text becomes the most comprehensive warning to human beings not to misuse the position which the Creator has given them."[123]

We have become the caregivers of the material world, whether things be in their natural state or crafted by human hands. Given our extraordinary reach, all of life now depends on us.

"Do Not Destroy": An Ancient Commandment for the Twenty-First Century

Bal tashchit, to be sure, does not mean no destruction at all. Rather, it urges us to consider how and why we use what we do. If the ways we use nature tend "to preserve the integrity, stability and beauty of the biotic community," as the pioneering American environmentalist Aldo Leopold taught, then it is right. And if the ways we use nature are compatible with the permanence of life, as Hans Jonas taught, and if our actions allow us to meet our needs today while enabling future generations to meet their needs tomorrow, as the United Nations put it, then it is wise use and fair.[124]

For we cannot live without using earth's gifts. To grow, to be, is to lay claim to other forms of life. We cannot build houses, make clothes, feed the world, conduct industry without disrupting the elements of nature. Consumption is necessary for existing. Raptors eat foxes, foxes eat squirrels, squirrels eat nuts. The Torah itself allows non-fruit trees to be cut down to provide the resources needed by the besieging army. Life is consumed to support other life. "[The Creator], in His infinite wisdom, ordained [life's] mutual interdependence in order that each individual being might contribute, whether much or little, to the preservation of the All."[125] We are all here as members of the whole, as both takers and givers to life's majestic enterprise. As humans, we cannot deny our impact on the world. Rather just the opposite. Genesis 2 asks us to recognize the scope of our power and manage it justly and well on behalf of the whole.

Maimonides, the great legal codifier and categorizer, sought to tease apart and clarify acceptable and forbidden acts of destruction. In the process he offered a new term: "Anyone at any time who chops down a fruit tree for *destructive purposes* [*derech hashchatah*] is punished. And not only regarding trees, but even one who breaks vessels or rips up clothing or tears down a building."[126] If we must destroy, and often we must, Maimonides is telling us to be mindful of how, when, and why we are doing it. Sometimes we must chop down trees. But when we do, it needs to be for a good purpose and in a minimalist fashion, disturbing the fewest resources necessary.

If a fruit tree is healthy, productive, and doing no harm, cutting it down would be forbidden. But, Maimonides continues, "if the fruit tree is damaging other trees or damaging another's field, or only produces a small yield and is not worth the effort to be maintained, it can be cut down." Once an object's utility ceases (see below for the conundrum of determining utility) or dwindles so that it demands more resources than its utility merits, or once the object threatens the well-being of others, then the object loses its claim of protection. And that makes sense. *Bal tashchit* seeks to maximize the productivity and fertility of the earth for the benefit of all. If there are elements of nature that themselves seem to obstruct or harm that productivity or fertility, like certain invasive species (both plants and animals) that engulf and destroy local biomes, those elements may be removed. Yet, we must proceed with caution when tampering with nature. All too often modernity has moved ahead in ways we thought were maximizing health and progress but which ended up harming the earth, the biosphere, and ourselves.

Still, we are permitted to "destroy" when such destruction contributes to the renewal of the earth. Such use should not even be considered destruction. It is rather generative, part of life's enduring cycles. "If the destruction is necessary," Hirsch tells us, "for a higher and more worthy aim, then it ceases to be destruction and itself becomes wise creating."[127] Our task is to be wise creators.

The devil of course is in the details. How do we determine what is essential and what is not, what is destructive and what is creative? Who is included in that calibration? How do we know what is enough and what is too much? How can we know what tomorrow's generations might need? How can we best anticipate the needs of our children in a future of a changed climate that lies beyond our science and imagination? In truth, our powers of prophecy are severely limited. But we can do the best we are able by constantly expanding the state of our knowledge, widening the circle of those who take part in these considerations, including especially those who are most affected by the decisions being made (such as those who live near power plants, data centers, highways, industrial sites, concentrated animal feeding operations). We also need to find ways to listen to voices representing future generations and consider what the imperative of intergenerational equity asks of us, as the phrase *m'dor l'dor*, "from generation to generation," urges.[128]

If the "destruction" we are now pursuing offers essential benefits for both today and tomorrow, if it supports the capacity of the earth to renew itself in beneficial ways in a meaningful amount of time, if it proves to be regenerative for future generations or at the very

How can we be mindful of future generations when, by definition, they do not have a presence or a voice at the table? How can we properly balance their claims of tomorrow against our claims of today? What metrics and bodies of authority do we need that can assess, implement, and enforce protections of the rights of future generations? Lawyers, ethicists, and economists across the globe are working on guidelines and formulas that allow us to answer these questions.[129] It is up to all of us to learn about them and integrate them into our various governance principles.

An externality is "a side effect or consequence of an industrial or commercial activity that affects other parties without this being reflected in the cost of the goods or services involved."[131] Some externalities are beneficial, like the benefits to a neighborhood when a coffee shop moves into a formerly blighted building or a park is created. Other externalities are harmful, like carbon emissions from fossil fuels or polluted stormwater runoff from impervious surfaces. Either way, the presence of externalities as part and parcel of the commercial activity is often ignored, with potentially harmful consequences.

least does not limit the regenerativity they depend on, then it could be deemed creative destruction.

There is one element of the classic law of *bal tashchit* that needs close examination. It is the assertion that one may cut down a healthy fruit tree if a good price can be realized from the sale of its lumber[130] or if one wants to build on the land where the fruit tree is (or by extension any other discretionary use that the presence of the fruit tree is preventing). Yet this interpretation comes perilously close to countenancing the short-term mindset of modern corporations, justifying decisions based on the limited interest of a few regardless of the loss of long-term benefits to many, including the earth.

The underlying value of *bal tashchit* encourages us to ask here: How is the overall value of the "tree," that is, the objects or area targeted for destruction, being assessed? Is the desire for money overriding concerns for environmental health? Whose interests are being considered in this decision? Whose are being ignored? What time frame is being used to assess financial value and return on investment? Does an immediate financial gain from the wood have a right to outweigh the long-term financial gains from the tree? What is being lost or harmed if the tree is cut down? Has there been, in modern terms, an environmental impact assessment on the loss of the tree? Has there been an assessment of the cumulative impact of this particular act on this particular neighborhood? And neighborhoods beyond?

There are times when Jewish law, as all law, needs amending, updating. This interpretation of *bal tashchit* is one of them. In his 1971 book *The Unforeseen Wilderness*, Wendell Berry, true to form, offered a pithy way to assess the wisdom of our environmental decisions and help us better understand how to assess the value of our deeds against the earth: "Do unto those downstream as you would have those upstream do unto you." If we would not mind being downstream in both time and place of the actions we ourselves are now pursuing, if we are not sacrificing certain neighborhoods for the benefit of others, if we can imagine those impacted in years to come thanking us for making these decisions, then such actions would merit being called wise use.

Environmental justice demands nothing less. So too does intergenerational equity. As economist Herman Daly wrote, "To hand back to God the gift of Creation in a degraded state capable of supporting less life, less abundantly and for a shorter future is surely a sin."[132] And so is passing forward a degraded world to future generations.

The mitzvah of *bal tashchit* teaches that we are not the measure of all things. For us to believe that all nature is subject to our capricious whims, that our desires can determine our actions without concern for the consequences, is a level of hubris the Torah does not endorse. All life makes claims. Not equal claims, not irrefutable claims, not absolute claims, but claims nonetheless. Balancing the claims of today and tomorrow, our generation and those yet to come, the big and the little, the urgent and the desired, the must from the maybe, is the challenge *bal tashchit* lays at our feet.

We have our work cut out for us. Our current patterns of consumption are destroying the ten-thousand-year Holocene era in which all of human civilization has blossomed and thrived. We know so much and yet still know so little. There is much we can learn from earth itself. With its four and a half billion years of experience, earth has had time to build the successful intricacies of nature. She can wrest energy from the rays of the sun, extract water soundlessly from the air, give every living thing their niche so that all thrive, contributing to the greater, interdependent whole.

This teaches us how the well-being of one aspect of nature is intimately tied to the well-being of the whole. We are learning how a mother tree passes along nutrients to the young trees in her orbit through a complex network of underground pathways, nurturing and aiding the forest's continued growth.[133] We are learning how reintroducing herds into grasslands and using rotational grazing can reverse desertification. We are learning how keystone species, no matter how big or small, are critical to keeping their patch of earth healthy. We must continue learning so that in our haste, hubris, or even honest efforts we do not harm nature. For nature will let us know (tragically sometimes too late) if we have overstepped our bounds.

"When a fruit tree is cut down," the rabbis taught, "its cry goes out from one end of the world to the other, but no one hears it."[134] Though we may be deaf to the cry, the pain is there. And so are the repercussions. When we violate the laws of regenerativity, when we destroy the fruits that hold the seeds of tomorrow, when our acts threaten our own well-being, the earth will cry out. And we need to listen.

Eight Guidelines for Ethical Consumption

Despite what we have been told about this being a dog-eat-dog world, life writ large is not me against you. In the words of German biologist and philosopher Andreas Weber, "The principles made apparent by biological research show us that life is, at nearly every level, a collective concern, a shared enterprise."[135] Pursuing the mutual needs of humanity in concert with the particular needs of nature is the way of *yishuv ha'olam*. It is not a matter of humanity versus nature, as if we were somehow distinct entities. We are, as Mary Oliver reminds us, one another's destiny.

Sefer Hachinuch is a sixteenth-century book from Spain which presents the 613 commandments in the order in which they appear in the Torah. Its authorship is uncertain.

Attending to the call of *bal tashchit* allows us to honor that destiny and hone the kind of people we can become. The author of *Sefer Hachinuch* writes that the prohibition of *bal tashchit* is not just an act but a spiritual discipline that "trains our souls to love what is good and beneficial and to cling to it." Living a life guided by this ethic sensitizes a person, he says, so that they will be "distressed at every waste and spoilage they see; and if they are able to do any rescuing, they will save anything from destruction, with all their power."[136]

Being alert to the impact of our consumption allows us to pause in our otherwise boundless pursuit of more. It humbles us and reminds us that the effects of our actions and our purchases ripple across space and time, that we are not independent moral sovereigns lacking obligations toward—or impact on—the physical world around us. Our task, then, is to do the right thing and in the process strengthen the earth's capacity to do what it is able to do: continually bring forth a life-filled world.

We can sum up Hirsch's guidance this way:

1. Do not take, use, buy, or make more than is needed.
2. Do not use more than is necessary to get the job done.[137]
3. Do not destroy what should only be used.[138]
4. Do not use, and potentially destroy, an object of higher utility when an object of lesser utility would suffice.[139]

Imagine creating a chart to help assess how well we match Hirsch's admonitions. Place the instructions from this list along the left side and the days of the week along the top. And for one week, at the end of each day, for every item we bought, every consumable we used, we assess how well we met the guidance on this list. At the end of the week we could review what we did, see where we shone, note where we could do better, and then start this exercise in awareness all over again.

	SUN	MON	TUE	WEDS	THURS	FRI	SAT
Do not take, use, buy, or make more than is needed.							
Do not use more than is necessary to get the job done.							
Do not destroy what should only be used.							
Do not use, and potentially destroy, an object of higher utility when an object of lesser utility would suffice.							
Do not consume more resources than the created object warrants.							
Do not destroy something out of whim.							
Do not waste an object's potential.							
Do not destroy an object's utility needlessly, "corrupting" it for future use.							

5. Do not consume more resources than the created object warrants.[140]
6. Do not destroy something out of whim.
7. Do not waste an object's potential.[141]
8. Do not destroy an object's utility needlessly, "corrupting" it for future use.

The mitzvah of *bal tashchit* speaks most directly (though not exclusively) to those who are blessed with material comfort. For those living on the edge, however, whether in our own cities or thousands of miles away, for those desperate to find ways to shelter and feed and care for themselves and their families, the worries of today trump the worries of tomorrow. Many in need cut down forests, releasing massive amounts of greenhouse gases, harming the land, and reducing biodiversity to raise cattle, or feed for cattle, or palm oil to feed first-world appetites. That degradation is on us.[142] We cannot countenance sending our environmental degradation offshore. Those of us with relative security must become aware of the impact of our consumer habits and work to craft and pursue systems and ways of consumption that do not harm anyone anywhere.

In 1903, speaking at the Grand Canyon, no less a leading light than Theodore Roosevelt urged all Americans to be guardians of the future and not believe we can plunder nature without harming the regenerative powers of the earth: "We have gotten past the stage, my fellow citizens, when we are to be pardoned if we treat any part of our country as something to be skinned for two or three years for the use of the present generation, whether it is the forest, the water, the scenery. Whatever it is, handle it so that your children's children will get the benefit of it."[143] What Roosevelt

Rage rooms offer patrons the opportunity to pay to violently destroy random objects just for the pleasure of destroying them—and vicariously, the source of their rage. The mental health efficacy of such exercises, however, is questionable, and there is some indication that it doesn't mitigate rage but feeds it.

In 2015, the United Nations adopted what it calls the 2030 Agenda for Sustainable Development. It includes seventeen goals that are to serve as "a shared blueprint for peace and prosperity for people and the planet, now and into the future." They include aspirations for no poverty, zero hunger, good health and well-being, quality education, gender equality, clean water, clean energy, economic growth, climate action, and more.

said of America we now know applies to all the world, especially as the international economy has become a grand village marketplace.

It All Matters

Some might argue that the morality of the destruction is contingent on the current state of affairs. If destroying something does not alter the well-being of the whole, if it causes no greater harm than its own loss, if there is more where that came from, then all is fine. Such destruction cannot be labeled wrong.

Bal tashchit teaches differently. *Bal tashchit* teaches that we are not to destroy wantonly—even if the whole is not harmed thereby. We may not shred a usable garment no matter how extensive our wardrobe is. We should not waste food even if we have a whole freezer full. We are forbidden to randomly chop down a fruit tree, or any tree now for that matter, even if we are in possession of a whole forest. We are to observe *bal tashchit* for the simple reason that in the language of the Jewish tradition, all things belong to God—and ecologically because all things emanate from and in turn must contribute to the generosity of this blue planet. We are mere transient beneficiaries with limited user rights. Other forces across the fullness of time gave of themselves so that we might have and use all that we do. We must keep that chain of generosity going.

"Holiness," the Rav, Rabbi Joseph Dov Soloveitchik, reminds us, "does not wink at us from 'beyond' like some mysterious star... but appears in our actual, very real lives... through the holiness of the concrete. An individual does not become holy through mystical adhesion to the absolute nor through mysterious union with the infinite... but, rather, through his whole biological life."[144] By using the matter of the world with reverence, by caring for all of Creation, by not wasting, we acknowledge the holiness that abides all around us.

An Emerging Guide to Commandments Between Humans and the Earth

Bal tashchit, then, is not just an environmental discipline; it is a spiritual one, enabling us to tread lightly and humbly on this earth. It seems time, then, to add that third category of environmental mitzvot (which we mentioned earlier) to the ethical and ritual categories the rabbis have divided the 613 commandments into. In addition to the commandments that define our behavior toward one another (*mitzvot bein adam l'chaveiro*) and the commandments that define our behavior toward God (*mitzvot bein adam laMakom*), we need to define our behavior toward the natural world (*mitzvot bein adam l'adamah*).

Below is an effort to flesh out this category so it may readily be translated into action. Blending the guidance of Hirsch with the United Nations' seventeen sustainability development goals,[145] along with the environmental justice community's seventeen principles,[146] I have assembled a list of principles that might inspire and guide us as we seek our enduring way on this emerald and blue planet.[147]

1. We are all guests on this rare planet earth, here but for a short while.[148]
2. It is with a visitor's gratitude and humility that we use the earth's resources.[149]
3. We are the current link in a sacred intergenerational covenant.

The seventeen principles of environmental justice were created at the 1991 multinational People of Color Environmental Leadership Summit and include "the sacredness of Mother Earth... that public policy be based on mutual respect and justice for all peoples, free from any form of discrimination or bias, mandates the right to ethical, balanced and responsible uses of land and renewable resources in the interest of a sustainable planet for humans and other living things" among others. I built this list inspired by these seventeen principles, the United Nation's 2030 Agenda for Sustainable Development, and Hirsch. And because Judaism favors the number eighteen—for it denotes life—I added one more.

4. We are the beneficiaries of the world left us by generations past.
5. We are the caretakers of the earth for generations yet to come.
6. Humanity is not the sole measure of Creation's value.
7. Proper use of a domain[150] must not diminish earth's regenerative properties.
8. Proper use must be assessed over a meaningful period of time.
9. Proper use of a domain must not deprive others of their fair share of earth's resources.
10. Proper use of a domain must not contribute to the degradation of earth's operating systems—large or small.
11. Proper use of a domain must not disproportionately harm select areas, neighborhoods, or populations, especially while benefiting others.
12. Those involved or affected should be part of the decision-making.
13. We should use no more resources than are necessary.[151]
14. As nature knows no waste, neither should we.[152]
15. We should abide in enoughness.
16. The urgency of "now" does not override all needs of tomorrow.
17. The production and use of earth's resources should be pursued in a cyclical manner: the end of one object's usable life should be the beginning of another's.
18. It is good to express our gratitude and offer thanks when witnessing or using earth's gifts and goodness.[153]

CHAPTER SIX

Yours, Mine, and God's

A refrain that reverberates throughout this book is that no matter how the universe came into being and who or what was responsible for it, we all can agree on one thing: it wasn't us. We did not form it, conjure it up, or in any way participate in its creation. It is here, as are we, born of a great mystery. This leads us to the age-old question: By what right and in what ways may we use the gifts of the earth?

The laws of *bal tashchit* answer one part of the question by teaching us how we are to *treat* the earth. But what are the laws and values that answer the other part of the question: How shall we *share* the earth?

How much of the earth can each of us justly claim as ours, and ours alone? Each of us, after all, deserves and needs our fair share. We need a place to live, a place we can call home, with doors we can close and chairs and a bed that wait just for us. Beyond a place of our own, as material beings we need to consume the gifts of the earth. Yet how much of earth's space and resources can and should we be able to call our own?[154]

Dozens of biblical commandments seek to respond to this question. There is the obligation of the farmer to bring animals, grains, and fruits to the Temple as sacrifices, sharing them with the priests, thanking God for these gifts. This allows the farmer to use the rest. There is the obligation to give tithes to the priests and the Levites—who are themselves landless—in return for their service in the Temple. There is the obligation to give tithes to the poor, as well as leave portions of one's field for them to glean, thus sharing the bounty of one's labor. All of these constantly reminded the Israelites to whom the earth belongs and framed their relationship to the land.

While the Torah does not state why *leket* and *shich'chah* are to be forfeited to the poor, the reason might be something like this: They both indicate that the farmer has more than enough. If our arms are so full that bundles tumble out of our arms, that might mean we are grabbing too much. And if we leave the field, forgetting sheaves when we return home, that might reflect that we have enough, for a poor or hungry person would not leave their food or livelihood behind.

But there is another astounding law, one that provides a truly radical vision of how we are to share the earth and conduct our ways of being with it. It is the law of the sabbatical year, the year called *sh'mitah*.

The Radical Subversiveness of Letting the Earth Be

Agriculture was the cornerstone of the ancient Israelite economy. Farmers were seen, in the biblical imagination, as caretakers of God's earth, managing the land's bounty for everyone around them: for themselves, the strangers, the widows, the poor, the priests and the Levites, and anyone else who had no land of their own. The biblical farmers knew, even as they sat in the light of their winter fires, planning the designs of their spring planting, that a portion of their future harvest would never be theirs. From the moment they set the bounds of their field and struck their plow in the warming earth, they knew that some of their harvest would belong to others. All that grew in the corners of their fields (*pei'ah*) would be reserved for the needy to harvest themselves. Any stalks the farmers dropped (*leket*) as they gathered in the produce of the rest of their fields would be forfeited to the hungry.[155] So would any sheath bundled and forgotten in the field (*shich'chah*). The farmers could not go back and collect them.

Such relinquishing of these claims was considered a matter not of "taking" from the farmer but of rightfully sharing God's bounty with others. Fertile earth, sufficient rain, the light of the sun, a place unravaged by war or pestilence, boundaries that are honored, the health of the farmer—all contribute to an abundant harvest. Success is a partnership involving God, farmer, land, and a peaceful society, so all, in some way, have a claim on its yield.

Such was the practice of Israelite farming in biblical times, year in and year out for a period of six years. Then, every seventh year, the rules changed. For every seventh year (*sh'vi'it*) was a year of rest and renewal for farmer and field, a time to radically reset society's structure and recast life as usual. Commonly called *sh'mitah*, the year of letting go, the seventh year was a time to reimagine not just the use and control of the land but the economy as well; a year that challenged assumptions about wealth and ways of being together. It was a year that called for letting all farm land lie fallow, but it was ever so much more than that.

Sh'mitah was to years as Shabbat is to weekdays, a taste of an ideal world, when everyone has enough and no one is in need. It was a time when our relationship to the world was neither consumer nor producer but celebrant. Every seventh year, landholders were commanded to release their rights to and control over their land. Fences were torn down, boundaries erased. The land of Israel reverted back to its native status, a bit of wildness, unowned, unworked, the commons, accessible, by God-given right, equally to all. Every *sh'mitah* year, the relationship of the people to the land, to its produce, and to one another was renewed, reclaiming the original promise of Creation. Leviticus speaks of the commandment this way:

> The Eternal spoke to Moses on Mount Sinai, saying: Speak to the Israelite people and say to them: When you enter the land that I assign to you, the land shall observe a sabbath of the Eternal. Six years you may sow your field and six years you may prune your vineyard and gather in the yield. But in the seventh year the land

While Leviticus focuses on the land and those who work the land, Exodus, more succinctly, emphasizes the poor: "Six years you shall sow your land and gather in its yield; but in the seventh you shall let it rest and lie fallow. Let the needy among your people eat of it, and what they leave let the wild beasts eat. You shall do the same with your vineyards and your olive groves" (Exodus 23:10–11).

> shall have a sabbath of complete rest, a sabbath of God. You shall not sow your field or prune your vineyard. You shall not reap the aftergrowth of your harvest or gather the grapes of your untrimmed vines; it shall be a year of complete rest for the land. But you may eat whatever the land during its sabbath will produce—you, your male and female slaves, the hired and bound laborers who live with you, and your cattle and the beasts in your land may eat all its yield. (Leviticus 25:1–7)

Once every seven years, the land, as a partner in God's divine covenant, enjoys a sabbath of rest. And everyone in the farmer's household shall honor that and enjoy the land's gifts. During this one year in seven, everyone is to be equal in their relationship to the land.

For six years life would be conducted as normal, guided by the standard expectations of commerce, property, appetites, and rituals that bound the Israelites to the land and one another. Some fortunes would rise; some would fall. Some would become rich; others poor. Some years would be bountiful, some lean. Across the six years, a sense of the individual would prevail: the land I work, the produce I grow, the goods I buy, the things I control. But every seven years that changed. Landholders no longer could claim their land; the poor were no longer restricted to the corners of the field and leftovers of the harvest. The land belonged to all—or better, to no one at all because it belonged to God. As Leviticus states, "The land is Mine; you are but strangers resident with Me" (Leviticus 25:23).

Every seven years, then, the standard rules guiding farming, ownership, and commerce were suspended.

Actively working the land was forbidden. Sowing the fields was forbidden. Harvesting the food that grew on its own in the normal way was forbidden. Treating the produce as commodities for sale was forbidden. Inappropriately disposing of produce was forbidden. The land and its yield were open equally to everyone, its goodness to be enjoyed by all.

The land was to be left alone. Unworked, the land would bring forth what it wanted, when it wanted. Everyone would come as equals and glean what they needed (a remembrance of the gathering of manna), even the land's once-and-future farmer would be a guest on his former land. Hoarding was not allowed. If you had a pantry full of carrots, then once the carrots were gone from the field, you would have to open your storehouse of carrots to the public. *Sh'mitah* was a year of connection not competition; a time of ours, and God's, but not any one's "mine."

It is not correct to think of the farmers as opening their fields or generously allowing others to glean as they would in the other years. Their land during *sh'mitah* no longer belonged to them. It belonged to no one. It became—once again as it had been at the beginning—the commons. Erstwhile "ownership" and user rights were undone. *Sh'mitah*, after all, means "to release, to let go, to relinquish," as when opening a hand or a grip on something you are holding.

If all law is an attempt to solve a real or anticipated problem, the question becomes, what problem is *sh'mitah* coming to solve? And could we imagine doing something of such magnitude in response to the many economic and social challenges of today?

Sh'mitah was a radical effort to provide a yearlong practical corrective to an economy and society that seemed destined to become less equitable, more stratified, less socially and economically mobile over time. The seventh year was a time to reboot. *Everyone* was to enjoy the same relationship to the land and its bounty. *Everyone* would enjoy the same land rights. *Everyone* would be called to the same tasks: wake, go to the local fields, collect their daily food, chat with their neighbors, share news. *Everyone* would experience a bit of the insecurity of the stranger, of the landless. There would be no rich, no poor, no them, only us. *Everyone* would see the humanity that all shared.

Millennia before some of today's governments began experimenting with "guaranteed universal income" to ensure economic security without the stigma of neediness, the Torah stepped into the breach. *Sh'mitah*'s actions—everyone harvesting the same fields the same way under the same restrictions—made it a social as well as economic leveler. Even at the end of the seventh year, when everyone reverted to their prior stations, the memory of gleaning in the same fields, of seeing one's neighbors bending over their baskets like supplicants all equally dependent upon their own labor and the goodness of the earth, must have left an enduring image.

And even if not, everyone knew that in another six years, the seventh year would come around and everyone would have to do this all over again.

A Year of Letting the Earth Be: A Commandment Called Sh'mitah, the Sabbatical Year

The very idea of *sh'mitah* is astonishing. And not just to us today. Even to the ancient Israelites, *sh'mitah* sounded outlandish, unrealistic. For the Torah hastens to tell us that immediately upon revealing this law, God anticipated the people's fear and resistance and quickly moved to reassure them. "You may wonder: 'What will we eat in the seventh year if we do not plant or harvest our crops?'" And God answers, "I will send you such a blessing in the sixth year that the land will yield enough for three years. While you plant during the eighth year, you will eat from the old crop and will continue to eat from it until the harvest of the ninth year comes in" (Leviticus 25:20–22). The Israelites were to be reassured that they could rely on the power and largesse of God. It was a reminder that in the biblical imagination, God was Creator, the rightful owner and manager of the land, able to make miracles. God would make the *sh'mitah* year work too. The pursuit of such a dream was so important that God would intervene and ensure that a socially and economically just society could be practiced at least once every seven years.

For the Israelites, then, *sh'mitah* was more than a year of agricultural husbandry, more than a year of rest for the land. It was a year in which the Jewish people were to remember to whom they and the land belonged, and what that called them to be. *Sh'mitah* invited all the Israelites to pause and review the ways they ran all their affairs. It was a time to reimagine their relationship to property, goods, wealth, one another, and the very structure of society. It was a year in which the inequities of normal life were to be cast aside and the fundamentals of an equitable life pursued. Even more than allowing the Jewish people to imagine such a world, *sh'mitah* afforded them the opportunity to practice it, experience it, with the hopes that they would carry bits of that experience into the six work years to follow. Over time, perhaps, the values of *sh'mitah* would come to guide the ways of the world.

In a way we would never have chosen, and in a way we could hardly have imagined, the first year of COVID-19 conjured up elements of a *sh'mitah* year. It forced a year of introspection, of changed productivity and consumer patterns; a year in which many of us reevaluated our lives, our work, our relationships. COVID-19 reminded us of how dependent we are on a healthy world and healthy neighbors; how united we are in our shared vulnerability (though some of us were more vulnerable than others). It reminded us that there are indeed limits to our use of nature and cautioned us against tampering with zoonotic boundaries. And at this remove, four years after global shutdown, we are already forgetting, which is why we need to continually relearn the lessons—and why *sh'mitah* returned every seven years.

Gerald Blidstein, a twentieth-century Israeli scholar who wrote a pioneering article on the subject, puts it this way: *Sh'mitah*, he says, was "more than a commonplace struggle between a radical religious demand and an unconsenting world. Rather, we have here an institution that in its essence contests the legitimacy of that world."[156] Of course the idea of *sh'mitah* was resisted, he argues. Who would choose such an upheaval? Who would want to give up land and power they believe are rightfully theirs? But that, he argues, is precisely the point. *Sh'mitah* was meant to disrupt the norm, to shine a bright light on economic injustices that the world had become accustomed to; to reveal an attitude of entitlement the privileged were otherwise blinded to; and to remind everyone to whom the land ultimately belonged. *Sh'mitah*, in this reading, was not just meant to offer a temporary corrective to the farmer's appropriation of the land and to inequality among people. It was meant, in its purist state, to urge society to find ways to imagine and craft an enduring society of sustainability, equity, and justice.

Yet, despite this hope, the authors of the Torah knew their audience.[157] They knew that people will be people, that the human spirit will likely continue to be fueled by the pursuit of status, acquisition, stratification, accumulation, ownership, and consumption as it always has. So the success of *sh'mitah* was not to be measured by its ability to stamp out such desires, for it would utterly fail if so. (And tradition knew, some level of desire and ambition is essential for the human enterprise to unfold. Else, the rabbis noted, our existence would be pointless, for we would be as boring, as compliant, as predictable, and as static as the angels.)

Rather, *sh'mitah*'s success would be to cyclically offer

an experience of a society that tempers these desires, tames them, restrains them, so that the human impulses of cooperation, justice, sharing, and enoughness could have a moment to prevail. The year of *sh'mitah*, through its radical design, serves to remind us of the good we can enjoy with the peaceful, equitable cooperation of the whole; that we are crucially dependent on the good earth and its continued health; and that we are responsible for protecting it and a just society for the welfare of all.

But there is something even more remarkable about the laws of *sh'mitah*. If one believes that the Torah was given by God, then the call to engage in such a recurring practice of equity and environmental care is not surprising. One would expect the Creator of all to preference a divine economy that uses well the resources of the earth on behalf of all humans and other creatures over the cumulative, inequitable stratification of an economy that overtaxes the capacity of the earth and creates classes that harden into castes. But if one believes that the Torah is a product of the human imagination, then *sh'mitah* is stunning, for it is something our ancestors fashioned for themselves (and repeated in Exodus, Leviticus, and Deuteronomy)! Our ancestors promoted—and generations have preserved—a self-imposed, society-wide expression of generosity designed to undo, however temporarily, the inequities that property-based, profit-motivated economies inevitably create.

True, the practice of this idealistic command ebbed and flowed over the course of time. Throughout the centuries when Jews lived mostly in exile, *sh'mitah* became little more than a memory, like the sacrifices of old. For the Torah specifically says that *sh'mitah* was to

Even Adam Smith, whose theories of benign self-interest have become the mantra guiding modern markets and much of our other behavior, believed in the essential empathic nature of humankind. "However so selfish man may be supposed, there are evidently some principles of his nature which interest him in the fortune of others, and render their happiness necessary to him, though he derives nothing from it, except the pleasure of seeing it," so starts *The Theory of Moral Sentiments*, published in 1759, seventeen years before *The Wealth of Nations*.

be observed "when you enter the land that I give to you" (Leviticus 25:2). That is to say, it was interpreted to apply only to Jews within the bounds of the Land of Israel, so it had little traction for Jews in the Diaspora. But with the advent of the modern Zionist movement in the nineteenth century and the influx of Jews making *aliyah* to Israel, the observance of the mitzvah of *sh'mitah* was reborn. The problem was, no one really knew if *sh'mitah* was even still operative as a biblical law, and if it was, how could it be observed in a society and economy so different from the biblical world? How could a modern kibbutz be sustained for a whole year with no harvest and no agricultural sales? How could the larger economy that likewise depended on such sales survive? And how could urban folk survive when they lived far from the fields that they were to glean during the *sh'mitah* year (given that *sh'mitah* produce could not be bought and sold as produce was in other years)?

And even if *sh'mitah* is to be practiced in the modern Land of Israel (which some choose to do), which specific tasks should be forbidden and which permitted? If you can't plow and seed the land, can you still irrigate and weed it? Do the laws pertain to a kitchen garden in one's backyard? What boundaries should delineate the biblically designated lands within which the laws of *sh'mitah* must be observed, and which lands lie beyond that? Even more, given that most Israelis do not live agrarian lifestyles, should the spirit of *sh'mitah* be translated from the laws of an agrarian society into the practices of an industrial-, information-, and finance-based society, and if so, how? These and other questions have been highly debated as Jews have returned to the Land of Israel, and they remain highly debated today.

This was not the first time that the laws of *sh'mitah* were questioned. No less an authority than Rabbi Judah the Prince, the redactor of the Mishnah, suggested almost two thousand years ago that the biblical laws of *sh'mitah* had been overtaken by the exigencies of history. No longer were the twelve tribes sovereign on their own land, no longer were the economic conditions of its observance the same, no longer could the people depend on the God-promised profusion of the harvest of year six to carry them through the harvest of the eighth year. So no longer should *sh'mitah* be obligated. Given the people's fear about the economic hardship and food insecurity it could cause, better to rule that this observance is no longer mandatory. If someone wanted to take on *sh'mitah* as a personal discipline and expression of piety, that would be fine. But the requirement for all who lived on the biblical lands to observe *sh'mitah* should be lifted. Suffice it to say that this was controversial in the times of Rabbi Judah, and such a view remains controversial today, though it is still promulgated by some.

But as with so many commandments in an ever-evolving, three-thousand-year-old tradition, *sh'mitah* continues to inspire. Communities of Jews around the world are exploring the enduring lessons encoded in this ancient law and how they can help us right the excesses modernity has wrought.

The Sabbatical Year's Relevance Today

Just as the modern concept of sustainability encompasses three intersecting arenas—people, planet, and prosperity (or as some call them: equity, ecology, and economy)—so too does the concept of *sh'mitah*. It promotes caring for the land, promoting social equity, and easing economic burdens by forgiving debt. How can we unpack these teachings so they may be meaningfully practiced today?

Caring for the Land

Sh'mitah is most often perceived as a corrective measure to heal the land from six years of work, for it is assumed that farming naturally depletes the soil's resources. But there are those today who urge us to see *sh'mitah* not as reparatory of past harm but as a reward, a bonus, for six years of regenerative land stewardship. *Sh'mitah*, in this view, is seen as a time of ease and bounty (as the Torah suggested) and not insecurity. It can be a time when the earth brings forth abundance on its own.

Such advocates of *sh'mitah* teach that the land—if treated well over the six working years—would, on its own, continue to feed us in the seventh. Orchards and perennials would continue to flourish. Seeds shed from annuals of the prior year's harvest would germinate and grow on their own. Soil healthy from six years of wise use would leave the earth ripe for nourishing the plants. The intensive acts of industrial farming would cease, while the cycles of productivity would continue.

As the proponents of this vision explain, "Regenerative agriculture... not only 'does no harm' to the land but improves it, using technologies that regenerate and revitalize the soil and the environment." Crop rotation, grazing rotation, cover crops, agroforestry (integrating trees into farm and grazing practices) all contribute to the health of the land. And not only does this practice improve the fertility of the soil, it has additional benefits the Torah never imagined, such as helping to reverse climate change. By building up the richness of the soil's organic matter, the earth is not only healthier, it can more readily and fully retain water and sequester carbon, as it has done for millions of years.[158]

Modern industrial farming techniques, on the other hand, contribute to soil degradation. According to soil scientists, "at current rates of soil destruction (i.e., decarbonization, erosion, desertification, chemical pollution), within fifty years we will not only suffer serious damage to public health due to a qualitatively degraded food supply characterized by diminished nutrition and loss of important trace minerals, but we will literally no longer have enough arable topsoil to feed ourselves."[159] Healthy soil is the key to healthy land, healthy people, and a healthy economy. Caring well for the soil is caring for our farms and our farmers, as well as ourselves. Along with the expansion of regenerative agriculture on traditional farmland, we must investigate and expand the opportunities of regenerative agriculture in urban farms and orchards.

Promoting Social Equity

Land was a major currency of wealth in the biblical era. Travelers, sojourners, hired hands, those who lost their land due to unfortunate circumstances, bad judgment, or bad husbandry—all were beholden to those who still had theirs. Landowners were thus charged by the Torah with being society's benefactors. They were to open segments of their fields to the landless, allowing them to provide for their daily needs. Welcoming gleaners to one's field was an act of justice and equity, not just kindness. Still, it reinforced social stratification, emphasizing the gulf between benefactor and recipient.

During *sh'mitah*, however, all that changed. Social stratification based on land wealth evaporated. Everyone assumed the same station before God and the same posture toward the land. Wealth and poverty, and the pride or shame that might attend them, had no foothold here.

And the Torah did not stop there. For every forty-nine years, that is, every seven *sh'mitah* cycles,

Inspired by *sh'mitah*, many Jewish institutions are creating seven-year plans to build better systems and structures that will improve their capacity to pursue sustainable practices in energy, food, waste, land use, equity, and financial investments. They are working with their members to install electric vehicle charging stations in their parking lots, plant rain gardens to manage urban stormwater runoff, buy renewable energy to power their buildings, increase plant-based dishes in their celebrations, and compost their food waste. It will be exciting to see what they have accomplished by the next *sh'mitah* year. And then seven years hence, and seven years after that, and seven years after that, for as long as we are counting time.

there was to be a jubilee. Even more radical than *sh'mitah*, this fiftieth year was to be a total social and property reset. All the land of Israel was to revert back to its original tribal allocation. Everyone was dealt a new hand. Everyone could start fresh. Everyone could begin again. While it is not clear that the dream of the jubilee was ever realized, the idea of the jubilee speaks to the enduring impulse of the Jewish people to seek equity throughout the land.

Even today, one's relationship to the land is liable to contribute to one's sense of self and often the ability or lack thereof to exercise self-determination. Using *sh'mitah*-thinking and the drive toward social equity, we can look more deeply at our collective choices of land use and see the presence or lack of equity in the distribution of its costs and benefits. We can better investigate why highways, incinerators, dumps, power plants, and heavy industry are located where they are; who benefits most from their services and who suffers most from their pollution; who owns the farm lands and why; the social and economic differences between homeowners and renters; the legacy of redlining and discriminatory lending practices that continues to impact neighborhoods and individuals today.[160] We can begin to reimagine the locations, quantity, and availability of urban farms, green spaces, and community gardens—expanding urban folks' access to land.

Sh'mitah-thinking can urge us to explore creative ways of responding to the social needs of our time. That is the second prong (people) of the three-pronged *sh'mitah* vision.

Forgiving Debt

And then there is the way that *sh'mitah* dealt with debt. In an agrarian society, one's livelihood was largely determined by the quality of the land, the vagaries of the harvest, the farmer's talents in sowing, irrigation, soil management, timing, and the serendipitous behavior of the weather, pests, animals, and more. To lose one's crop—not at all unthinkable—was to lose one's livelihood. A loan could be the only way to get through the year. And if the harvest the next year was not enough to pay off the loan from the last, the debt would grow. The accumulation of failed harvests could lead to losing one's land, and in the biblical period it could literally require giving oneself over into servitude.[161]

The Torah sought to avoid this scourge of debt by embedding a solution in the *sh'mitah* year. "Every seventh year," it teaches, "you shall practice the forgiveness of debts. This shall be the nature of your forgiveness: every creditor shall cancel the debt that he claims from his fellow; he shall not dun his fellow or kinsman, for the remission proclaimed is of the Eternal" (Deuteronomy 15:1–2). In the divine calculus of right versus mercy, every seven years mercy wins. It is not easy for the lender to write off that wealth, the Torah acknowledges, but in the assessment of the divine economy, it would be merciful, even just. It was the lender, the Torah assumed, who could better absorb their loss than the borrower could endure their poverty.

The lenders knew that all such loans were unsecured and could vanish like vapor in the new year. The shorter the time between loan and payback, the greater the risk of the borrower defaulting. This must surely have dampened lenders' desire to offer loans. Indeed the Torah anticipated such lender reluctance here too, so it cautions: "Beware lest you harbor the base thought, 'The seventh year, the year of remission, is approaching,' so that you are mean and give nothing to your needy kin—they will cry out to the Eternal against you, and you will be held guilty" (Deuteronomy 15:9). Generosity and compassion must rule the day.[162] And if anyone should balk, it was God who would enforce compliance.

To be fair, this call for forgiveness of debt did not catch either the lender or the borrower unaware. An average Israelite always knew what year it was in the *sh'mitah* cycle, just as they knew what day it was in the week, for a simple reason: the recipients of the annual tithes were determined by the particular year of the *sh'mitah* cycle. Here is how it worked: The Levites were the one tribe of all the twelve that did not receive a patrimony when Joshua carved up the Land of Israel, for they were dedicated to the work of the Temple. Instead, they lived off the gifts of the people. One of those gifts was the annual tithe. Every first, second, fourth, and fifth year the people gave tithes directly to the Levites. The tithe of the third and sixth years, however, was given to the poor. The common folk, the Levites, and the poor all were well aware of the count of years and who would be the recipients of the tithing. And because in the seventh year debts would be erased, borrowers and lenders were both keenly aware of the count of the *sh'mitah* cycle, for both were eager to know how much time was left to repay the loan.

The biblical tradition of forgiving debt every seven years, of regularly providing for people's basic needs, and of ensuring that everyone gets a fair shot at life has inspired many to explore how we can overcome the financial ills of our society today. How, they wonder, can we blend environmental health, individual dignity, and economic stability? The structure of *sh'mitah* emboldens such experimentation: By default it is only one year. So take this idea, tradition seems to call to us, and run with it for a short while. Give it a try. If it doesn't work, come year-end the experiment ends. If it does work, find ways to continue, expand, and improve it.

There is no one way to adapt *sh'mitah* to our lives today. The academic world has offered year or half-year paid sabbaticals for over a century. Some clergy also enjoy weeks-long or month-long sabbaticals. Even businesses are now beginning to offer month-long, or more, sabbaticals to certain employees. We can encourage every entity we are involved with—ourselves, our families, our congregations, our businesses, the boards we are on, the groups we belong to, the banks we invest in—to ask these questions and embark creatively on a seven-year plan focused on the three pillars of *sh'mitah.*

A New Application of the Sabbatical Year: Embracing Rest for the Soul

The social vision of *sh'mitah* can inspire us to apply its lessons in another area of life: the personal.[166] As we walk through life, it is not just land and debts that we need to occasionally reevaluate and release. It is also the grudges, hurts, memories, disappointments, dreams we carry that no longer serve us well. Yet it is hard, sometimes frightening, to find the right moment, the spiritual strength, to release them. We often find comfort in our known emotional pain. It is familiar, it helps justify our actions, and in part it defines us. It is made of memories that have shaped who we are. But when those memories and feelings become anchors tying us to a past from which we need to be free or tread-

The benefits of canceling debts today could be significant. The Education Data Initiative estimates that canceling student debt in the United States may add $109 billion on average to the annual GDP for the next six years and add up to 1.5 million new jobs.[163] Debt relief can also have intergenerational impacts, allowing families to buy homes and accumulate a measure of wealth that can give a boost to their children's education and financial security. And the mental health benefits of being relieved from the worries of financial insolvency aids in individuals' physical health, promotes better decision-making, and allows individuals to engage more productively in work, family, and pleasure.[164] Debt relief for countries is also seen as a matter of justice as well as financial wisdom.[165]

mills that prevent our moving on, it may well be time to move beyond them.

We can take inspiration from the *sh'mitah* year to rid ourselves of this emotional debris. We can create rituals of release and divestment. We can select the object, the place, the moment of change. We can invite our friends to join us or do it alone. Like *tashlich*, the Rosh Hashanah ritual in which we symbolically toss our sins into the water, *sh'mitah* can inspire us to toss our unwanted past into the ether, unload a burden that has weighed us down, and allow us to emerge lighter, newer. We, of course, can do this any time, any year. But *sh'mitah* is a ready-made time for this, for recasting ourselves, becoming that person that we have said we want to be. It is a time that allows us to announce this to our friends and family as well, a time preordained for new beginnings.

Such lightening of our spiritual load can help us lighten our material load as well. Many of us likely own more stuff than we need—more than we even use. As the Israelites were bidden to open their well-stocked storage bins to the public during the *sh'mitah* year, so every seven years we too can open our well-stocked closets, attics, and storage units, explore what is in there, divest ourselves of things we don't need or no longer use, and donate, sell, or recycle items that can be of greater benefit to others.

This divesting is a gift we give ourselves.[167] For clutter, surprisingly, is not inert. It does not just sit there. It takes up space, presses in on us, demands our attention even if only to pay the monthly check to the storage unit. Being surrounded by clutter can add to our stress. Those geegaws, gadgets, and tchotchkes we bought to add joy to our lives may end up gathering dust and adding to our dissatisfaction.[168]

One year, when I working as the Jewish educator at the Baltimore Jewish Community Center, I took a small group to our local dump. (The county prefers to call it a landfill.) Everyone was instructed to bring something they had been holding onto for too long so they could finally, publicly, ceremoniously rid themselves of it. Some people brought letters and mementos of relationships long past, or bundles of legal papers no longer needed, or tchotchkes from prior eras of their lives. With a heave and a ho, they tossed these objects over the retaining wall and watched as they landed with a satisfying thud. While it might take more than a shove to rid oneself of all the debris of one's past, divesting of things in such a ritual manner offered some welcome relief. As we are more mindful now of the urgings not to waste, it would be better to arrange a gathering of these items and a ceremonial donation to Goodwill Industries or some other entity that can wisely use our castoffs.

This is not suggesting wholesale purging (unless it is indeed time for that). Rather it is to encourage the intentional culling of our unneeded, even counterproductive, possessions, affording us the welcome opportunity to reframe the narrative of our lives.[169] A friend of mine, determined to come to terms with her years-ago divorce, took a wedding gift in hand—a glass serving dish etched with his-and-her initials—stationed herself above her apartment house dumpster, and smashed the plate to smithereens. A threshold had been crossed. A new identity was born.

The *sh'mitah* year does not just suddenly end with the advent of Rosh Hashanah of the eighth year. It culminates in a special gathering of the people during the holiday of Sukkot of that eighth year: "And Moses instructed the people as follows: At the end of every seventh year, the year of *sh'mitah*, [in the eighth year] during Sukkot, the Feast of Booths, when all Israel comes to appear before the Eternal in the place that [God] will choose, you shall read this *torah* [Teaching] aloud in the presence of all Israel. Gather the people: men, women, children, and the strangers in your communities. So they may hear and so learn to revere the Eternal your God and to observe faithfully every word of this Teaching" (Deuteronomy 31:10–12).

As one *sh'mitah* cycle ends, then, another begins, with the gathering of the people—young and old, resident and strangers—in one place, at one time to hear the word of God. It is as if they were gathered again at Sinai, hearing how to be a good and just people for the first time. And as they return home after the holiday, as they return to the everyday laws of ownership and the marketplace, they are to take with them the teachings that undergird their society.

One of the more remarkable aspects of *sh'mitah* is that while it tells us what we cannot do during the seventh year, it does not tell us what we must do. It does not set a goal or stipulate how we are to fill the hours we no longer must spend on our livelihood. It offers only a hopeful, dream-filled space. The rest is up to us.

Rabbi Abraham Isaac Kook tells us that *sh'mitah* provides us with an "image of a world that is good, upright, and godly—aligned with peace, justice, grace, and courage, all filled with a pervasive divine perspective that rests in the spirit of the people."[170] May we too be inspired by the recurring vision and practice of *sh'mitah*, note it on our calendars, and anticipate its arrival so we may live more deeply into the goodness we seek and the earth we so desperately need.[171]

CHAPTER SEVEN

All We Need

Of all that exists, of all that is available to us, of all that we can rightfully claim as our own, the question that continually confronts us is: How much is enough?

Writing in the *Journal of Retailing* in 1955, marketing guru Victor Lebow offered his audience this advice: "Our enormously productive economy demands that we make consumption our way of life, that we convert the buying and use of goods into rituals, that we seek our spiritual satisfactions, our ego satisfactions, in consumption. The measure of social status, of social acceptance, of prestige, is now to be found in our consumptive patterns. The very meaning and significance of our lives today is expressed in consumptive terms."[172]

Lebow was serious. Writing with the cadence and urgency of a prophet calling people to faith, Lebow presented this heartfelt articulation of the new American economic ethic. With the constraints and deprivations of the Depression in the past, with the winds of victory lifting our spirits, with a wartime manufacturing industry all dressed up and still raring to go, American producers needed to find new markets and new products, and most of all, expanded consumer appetites, to keep their machinery humming. Manufacturers were eager to promote a "more-is-more," endless-growth mentality. Consumers were eager to purchase new, cheaper, and guilt-free products. Gratitude was found not in enough but in the anticipation of more. Citizens became consumers. Buying became our obligation as well as our passion. Never was self-indulgence more patriotic.

"We need things consumed, burned up, worn out, replaced, and discarded at an ever-increasing pace," Lebow continued. "We need to have people eat, drink, dress, ride, live, with ever more complicated and, therefore, constantly more expensive consumption." Not a thought was given, nor a word expressed, about the impact of such consumption on the environment or what such indulgences would mean for future generations—or, for that matter, the toll such ever-wanting might take on our souls.

So, for over half a century, Americans, and much of the Western world, bought into this fantasy. We delighted in going shopping. We believed endless growth was possible, desirable, and achievable. No need to worry about the call of equity or the limits of nature. Market forces would find a way to make things right. Even more, we were told that happiness—not its pursuit but its very attainment—was our birthright. It was not just what we deserved but what we were owed. Happiness was our birthright, to be found in our increased acquisition and accumulation of things. Lest we, the consumer, ever falter, ever doubt, or ever somehow merely forget to do our duty, "[we] will be gently and insistently reminded that [we have] no right not to be happy."[173]

And why not? Cheap energy powered our homes and cities. The earth opened itself to us, offering up its seemingly endless resources for us to mine and extract. The pollutants and waste we created were buried in distant places we designated "away" or were jettisoned into the air or dumped in the waters. We were either oblivious to or unconcerned about the consequences. Human ingenuity was busy developing technology that was rapidly bringing good things to life.[174] The sky was the limit. Or so it seemed.

It took until the 1960s, when Rachel Carson warned the world of the perils of DDT, when Vietnam vets and millions of acres of land were suffering from the effects of Agent Orange, when the Cuyahoga River caught fire, that the public began to wonder about the toll that

The Cuyahoga River Fire & the Founding of the EPA

On Sunday June 22, 1969, the Cuyahoga River in Cleveland, Ohio, caught fire, and not for the first time. The Cuyahoga was a river lined with industries that regularly dumped their detritus into its waters, covering it in thick swaths of oil. Everyone knew that the river was polluted. It was the price they were willing to pay for industry and "prosperity." The fire lasted thirty minutes, caused only a modest amount of damage, and garnered only a minimal amount of local coverage. It was not the only US river to catch fire. The Buffalo River (Buffalo, New York) erupted in flames the year before. The Rouge River in Detroit, Michigan, caught fire later that year. But as fate would have it, it was the Cuyahoga River fire that became an icon of America's river pollution and one of its greatest catalysts for spurring the nation to action. Within a year, the first Earth Day would be held, championed by Senator Gaylord Nelson and attracting twenty million participants nationwide. By the end of 1970, Congress, prompted by events, the national zeitgeist, and the leadership of President Richard Nixon, would create the Environmental Protection Agency.

Early Warnings of Global Warming

We have known about the devastating dangers of a warming earth for some time. Early warnings that greenhouse gases could alter the nature of the earth's atmosphere were sounded in 1856 by Eunice Newton Foote. In 1859, John Tyndal discovered the same thing. A century later, in 1959, the research of Charles David Keeling, a pioneer in measuring atmospheric levels of carbon dioxide, showed two things: the diurnal and seasonal variations of atmospheric carbon dioxide due to the earth's breathing in and out, and, presciently, "where data extend beyond one year, averages for the second year are higher than for the first year."[175] The atmosphere was heating up. This has been dramatically shown over the past decades through a graph referred to as the Keeling Curve.

In 1965, President Lyndon B. Johnson created his Science Advisory Committee and charged it with looking at air, land, and water pollution. The committee created a working group that concentrated on atmospheric carbon dioxide. As the American Chemical Society reports, "A subsection of the report [of the carbon dioxide working group] explored possible effects of increased atmospheric CO_2 on Earth's climate. These ranged from significant increases in average global temperatures, to melting of arctic ice sheets and the resulting rise in sea levels, to increased acidity of water bodies. The report concluded: 'Through his worldwide industrial civilization, Man is unwittingly conducting a vast geophysical experiment.' "[176]

our progress and our consumption was taking. And it took the courage of a children's author like Dr. Seuss to publish, in 1971, *The Lorax*, showing how the world's gluttonous appetite was leading to unrestrained environmental degradation.

Still, responding appropriately takes a lot longer. Redirecting a roaring economy from a linear, disposable model to a cyclical, reusable one and transforming an equally expectant public appetite for ever more into the pursuit of enough is not easy. It takes a spiritual revolution to pivot from consumer indulgence to the simpler delights of enoughness. Even more, those who seek such a pivot, who speak on behalf of temperance and enough, are told that without all our buying, without the churning of the market that our constant consumption provides, the economy will crumble.

Yet, we are increasingly realizing that we need to craft a world that works the other way around. It is the current market model with its chimera of endless growth that threatens to do us in. As the British economist Kate Raworth teaches, we need to "reimagine the shape of progress. Today, we have economies that need to grow whether or not they make us thrive, and what we need... are economies that make us thrive whether or not they grow."[177]

In the 1970s, Senator Gaylord Nelson from Wisconsin boldly stated the intimate connection between nature and the economy: "All economic activity is dependent upon the environment and its underlying resource base of forests, water, air, soil, and minerals. When the environment is finally forced to file for bankruptcy because its resource base has been polluted, degraded, dissipated, and irretrievably compromised, the economy goes into bankruptcy with it."[178]

This makes sustainability the Great Work of our generation for three reasons: the earth is finite, everything is interconnected, and more rarely makes us happier.

The Earth Is Finite

There are limits to earth's capacity. While humanity has long known that here and there the earth could be harmed by overuse and mistreatment, until the modern era it was assumed that the earth itself was eternally resilient. The planet seemed so big. We could imagine a town, a field, even a region being degraded, but surely not more. It was hard to imagine how humanity could possibly upset the ocean, the atmosphere, the mountains, the climate, even the tilt of the planet.[179] These were features of nature, occupying a realm far beyond the reach of humankind. If we, unwisely, harmed our patch of the earth, we could move, go elsewhere, and allow nature to employ its healing devices to restore what we have destroyed. Life would return. That was the promise of Genesis 1—life was created with seeds of eternal renewal, capable of infinite regenerativity beyond any destructive capacity of humankind. Or so we thought.

This premise, we now know, is no longer true.

The current pressures of human population and our growing appetite are increasingly destabilizing the earth's operating systems. We have filled the earth—as Genesis 1 bids us to do—overrun it really, with our desires, debris, and technology. Earth is struggling to support us.[180] For decades, our collective annual global consumption and concomitant creation of waste has exceeded the carrying capacity of the earth. In 1970, earth's "overshoot day," the day in the year when we begin consuming more of earth's resources than it can replenish annually and discarding more waste than it can absorb, was December 30, thus practically a world of parity. In 2023, though, overshoot day was August 2.[181] And the calendar gets shorter every year. If all the world consumed like the United States, the world overshoot day would be March 13.[182]

To continue with business as usual—extracting, producing, consuming, and discarding as we have—is not only unsustainable, it is immoral. For it harms not only us but all life that has perforce become dependent on us. All life yet to come. And the earth itself, the very source that has given us life and nurtured us.

The thirteenth-century sage Joseph ben Abraham Gikatilla put it this way: "To what can we compare those who reap enjoyment from God's world but do not take care of it? To a king who places his garden in the trust of a caretaker who in turn destroyed it and failed to protect it. So too, every person who eats and drinks and benefits from the world, but only attends to his own benefit and enjoyment, such a person destroys the world."[183]

Interconnectivity: There Is No Place Called "Away"

Everything is connected to everything else. What we dig up, use up, burn up, dump out, cover over, toss aside, and otherwise cast back to the earth, sea, or air does not disappear. Sooner or later it comes back to haunt us, in one form or another. And pollutants travel broadly. Microplastics—those tiny particles five millimeters or less that are intentionally added to our cosmetics, detergents, paints, and more, as well as the plastic coming from larger items that become brittle and fracture into myriads of tiny pieces—enter our air and waterways and eventually our bodies. They are in the foods we eat, the water we drink (even our bottled water[184]), and the air we breathe. They have been found in human blood in 80 percent of the people tested[185] and in every one of the sixty-two placental tissue samples tested, as well as in breast milk.[186] It is still unclear the impact they will have on our health, the health of our children, and the animal world that also ingests them, though they have been found to damage human cells in the lab.

Of course it is not just microplastics that get dispersed around the globe. Carbon dioxide, methane, nitrogen oxide, mercury, sulfur dioxide, larger particulate matter, and more are spewed into the air from burning fossil fuels. It is estimated that only 3 percent of the earth's ecosystems remain untouched by human activity.[187] We do not live in silos.[188] Personal and political boundaries do not correspond to nature's boundaries. Water and air do not stay put. They flow, affecting areas beyond their bounds. So what is mine is not fully and exclusively mine. What is yours is not fully and exclusively yours.

None of the air and water that passes through our bodies today is new and virginal. Over millions of years, it has passed through the bodies of our ancestors—human and not. In short, we imbibe and process the essences of one another. Everything eventually returns to the commons, those natural resources that we all share. And access to a healthy commons is the birthright of all. Even more, what we do today will also affect those who come long after we are gone, for those who follow will inherit the legacy of our stewardship, good or bad.

The rabbis taught the truth of this interconnectivity with this story:

We are mere stewards of land we manage, with rights of fair use but not ownership, as discussed in chapter 5, which explores the concept of *bal tashchit.* That is why the pious man said the farmer did not own his land but that he did "own" the public road or path onto which he threw the stones, for it is part of the commons, the area which everyone—including the farmer—shares and uses. In that moment, the farmer realized the wisdom of the one he had thought foolish. The harm of throwing stones on a public thoroughfare can readily be remedied, unlike much of the harm we are causing today.

> A person should not discard debris from their own property and deposit it on public property. Once, there was a certain farmer who was removing stones from his property and throwing them onto public property [i.e., a road or path]. A pious man saw this and said to him, "Foolish man, why do you remove stones from property that is not yours to property that is yours?" The farmer just laughed at him, thinking surely it was the pious man who was the fool.
>
> Some days later the farmer sold his field, and when he was walking on that public property [where he had thrown his stones], he stumbled over those stones. He then said, "How well did that pious man say to me, 'Why do you remove stones from property that is not yours to property that is yours?'"[189]

The earth is of one piece. There is no "away."

More Rarely Makes Us Happier

Study after study confirms that happiness is more a product of the spirit we possess than the goods we own. As Robert Waldinger, the director of one of the world's longest studies on well-being and happiness (more than eighty-plus years and still going), reports, the quality of our relationships trumps genes and wealth in predicting our quality of life and its longevity. "The people who were happiest, who stayed healthiest as they grew old, and who lived the longest were the people who had the warmest connections with other people."[190]

Indeed, the endless pursuit of objects of acquisition is often correlated with lower levels of happiness and joy. "Beyond the satisfaction of basic needs (e.g., food, shelter, and access to health care), increased consumption does not consistently deliver greater happiness.... Those individuals placing more importance on acquiring goods to improve their happiness and status have been shown to report lower life satisfaction."[191]

Despite Lebow and other misguided happiness hawkers, then, seeking ever more is like running on a treadmill. We never arrive at our final destination. Indeed it's not even clear what or where our destination is. We are just running. And that takes a toll on both body and soul. For at some point we need a place of rest, of ease, of arrival, of contentment. But constant desire, the urge for ever more, is a harsh taskmaster. Even when we have enough, we are not satisfied. So we continue to seek, and consume, and pursue happiness in a constant state of hunger and a wasteful striving for more.

Over one hundred years ago, Rabbi Samson Raphael Hirsch warned about the futility of wanting ever more: "For the one who does not set limits to unrestrained impulses, the universe itself—and eternity in the end—become too small to satisfy."[192] The Book of Ecclesiastes, written over two thousand years ago, presents ever wanting not just as a regrettable human impulse but as a curse: "The one who pleases God is given wisdom and knowledge and joy, while the one who displeases God is given the urge to gather and amass" (Ecclesiastes 2:26). Endless desire, the author suggests, is a Sisyphean punishment, a hunger that gives no rest.

Finding Happiness

It's not that we don't experience momentary highs with new things. We do. We finally get that latest gadget, those perfect shoes, that salary we pined for. And for a while, we are thrilled. We feel great. But over time

Shopping therapy is a real thing. Spending money and buying things does tend to release neurotransmitters, like dopamine, serotonin, oxytocin, and endorphins, so for a short while we experience reduced anxiety, less pain, enhanced feelings of ease and excitement. That is why we do it. The problem is, the feelings don't last. And when they are gone, all we have left is the bill and the object that we now have to find space for on our shelves or in our closets. There are better alternatives: "Nature, music, exercise, and bonding with loved ones can be therapy, too [releasing the same neurotransmitters]. These activities save you money and enhance your physical health and relationships."[193] So too can laughter, aromatherapy, becoming a regular attendee at morning or evening minyan (daily Jewish services), gardening, or volunteering. These alternatives are often free, reliably reproduceable, enduring, and healthier for oneself and the earth.

our euphoria dissipates; our baseline emotional state reemerges. We are back to the beginning, back to seeking more.

The hedonic treadmill, or hedonic adaptation, is the name given to this tendency to return to our baseline state of happiness (be it joyous or not) after experiencing a bump or dip due to external circumstances. While this is most helpful in response to negative events that happen in our lives, for it allows us to "bounce back" or return to daily life with a modicum of zeal, it is unfortunate in response to pleasurable events, for the joy of the moment fades and we are back to our baseline, whatever that might be. The challenge seems to be, then, not how to get as many hedonic bumps, but how to elevate our baseline state of happiness.

So, despite what the ad men of yesteryear and the commercials of today tell us, buying ever more is not the way to enduring happiness. Nor is it the formula for economic salvation. Nor the pathway to environmental well-being. Quite the contrary. Pope Francis speaks of this when he writes, "The external deserts in the world are growing because the internal deserts have become so vast."[194] Try as we might, we cannot rightly fill a spiritual need with material purchases. While buying things (and thus using more of earth's resources) may momentarily soothe our spiritual aches, in the end it is not the balm we need. We should not be seduced into denuding the world of its bounty to soothe our souls with material counterfeits.[195]

We all seek happiness. It is a universal impulse. We seek not just the euphoria of fleeting joy or the passing feelings of situational delight, but deep, enduring, soul-filling happiness. It is no wonder that the very first word of the Book of Psalms is *ashrei*, "happy are those. . . ." The question is, how do we fill that ellipsis? One answer that has echoed across the ages is gratitude.

Gratitude

Though the shape of our days is determined by the situations we find ourselves in, the quality of our lives is determined by how we respond. We are the curators, the meaning-makers, the storytellers of our lives. We are the ones who assign commentary to our experiences. We can choose to fill our days with grievances and grumpiness, or we can offer expressions of gratitude. Psychologists tell us there is great benefit to doing the latter. Gratitude, they say, helps people become more optimistic, more generous, less anxious, more resilient. Being grateful opens us up, broadens our vision, and expands the boundaries of self we sometimes draw too tightly around us. Gratitude allows us to recognize and appreciate gifts and kindnesses that come unexpectedly, often unearned. And it is an attitude that can be nurtured and grown.

Some people are just naturally grateful. Gratitude is their default way of being. Whether finding a convenient parking spot or recovering from surgery, they are grateful. They see life as a gift, not a promise. Others of us, perhaps most of us, need a bit of help exercising our feelings of gratitude. The good news is that the spirit is malleable, allowing us to become grateful through practice. Noting and naming the things that go well (or don't go badly), speaking or writing about them, saying thank you out loud more often, have all been shown to awaken gratitude. And when practiced, intentionally and regularly, gratitude can burrow deep inside, become a habit, our default orientation, accompanying us across life's many years, both challenging and joyous.[196]

What does the psalmist say is the source of happiness?

Happy is the one
who has not followed the
counsel of the wicked,
or taken the
path of sinners,
or joined the
company of the insolent;
rather, this one delights
in the Eternal's teaching
and studies that teaching
day and night.
Such a one is like a tree
planted beside
streams of water, which
yields its fruit in season,
whose foliage never fades,
and whatever
it produces thrives.
(Psalm 1:1–3)

Although constant study may not be your cup of tea, still we can learn from the psalmist that finding something of enduring value, something that binds us to a community of purpose, to something that is larger than ourselves, to a source of wisdom, that fills us up, that makes us feel alive, goes a long way to making us happy.

"Happiness," writes the great psychologist Mihaly Csikszentmihalyi, "is not the result of good fortune or random chance. It is not something that money can buy or power command. It does not depend on outside events but, rather, on how we interpret them. Happiness, in fact, is a condition that must be prepared for, cultivated, and defended privately by each person. People who learn to control inner experience will be able to determine the quality of their lives, which is as close as any of us can come to being happy."[197]

The British philosopher and poet G. K. Chesterton explains how simple dwelling-in-gratitude can be, and how uplifting:

> Here dies another day
> During which I have had eyes, ears, hands
> And the great world round me;
> And with tomorrow begins another.
> Why am I allowed two?[198]

It is in the expression of joy and amazement and the lack of feelings of entitlement that gratitude flourishes.

Two thousand years ago, a sage named Ben Zoma felt this same joy, amazement, and sense of gratitude at every-day life. Whenever Ben Zoma went to Jerusalem, we are told, he found himself witness to the city's creativity and industry. Throngs of people were milling about, selling their wares, offering lodging, filling the air with the din of life's daily vitality. And in response, he offered this prayer:

> Blessed is the master of the secret ways of the world who created all of this so I may benefit. What labors Adam was burdened with and how hard he had to work before he had even one morsel of bread to eat! He had to sow, plow, reap, bind, thresh, winnow, separate, grind, sift, knead and bake, and only then could he eat. Whereas I get up in the morning and find all these things arrayed before me. And what labors Adam was burdened with before he could don a [wool] garment! He had to shear, wash, separate, dye, spin and weave, and only then could he put it on. Whereas I get up in the morning and find all these things arrayed before me.[199]

Ben Zoma's response to this everyday world was a megadose of gratitude. Gratitude that he was able to benefit from all this. Gratitude for the people who did the work; gratitude for the ingenuity of the human mind that devised the skills and the crafts to produce such goods; gratitude for the society—organized and peaceful—that allowed the marketplace to function; gratitude for the health of the earth that brought forth all these resources. Gratitude for the power that held it all together.

Ben Zoma saw what others miss and came away enriched by more than the goods and services he could buy there. Witnessing the complexity of life, the details that meshed to make a smoothly running society, he was filled with awe, appreciation, and a deep measure of gratitude.

I imagine we rarely feel the same rush when we walk into the grocery store and see the shelves lined with food or when we shop online and see the plethora of options and ads—or even when we glance around our home and see all that we are privileged to have. Yet we too can, and should, wonder and marvel not only at the goods themselves, but at the people, the wisdom, the earth, and the systems that brought them into existence and into our hands.

Gratitude must have been a signature attribute of Ben Zoma, for of all the teachings he offered over the course of his life, his sayings about gratitude are quoted the most. In the two-thousand-year-old rabbinic collection *Pirkei Avot* (Ethics of the Fathers), he is remembered as saying, "Who is rich? Those who are grateful for their lot."[200] We are blessed, this saying teaches, if we assess the good we have and recognize it as the gift it is. Ben Zoma seems to want us to know that gratitude is not an all-or-nothing proposition. It is not a destination only to be appreciated when we finally and fully arrive. It is rather a companion that accompanies us along on our journey, a feeling we can have at every small something that serves us well.

By saying we should be grateful for our lot, Ben Zoma is not saying that we should settle for what we get or what we have. To be grateful is not a call for quietism, of accepting life as it is handed to us without complaint or desire or ambition. It is not a pacifying call to concede our lives to fate, come what may. Ben Zoma was too much of an adventurer and seeker to suggest that. His spirit was full of curiosity and exploration. He delighted in discovery. Rather his saying teaches that despite all of life's disappointments, despite all its gaps, gaffs, hurts, and imperfections, we can still be grateful for the good we have. For we are rich if we can find a bit of joy in the good we possess today without waiting for more good, which we still hope comes tomorrow.

Ben Zoma is also remembered as speaking of gratitude when speaking about hospitality: "What does a grateful guest say? 'May the host be remembered for good! Look how many wines he brought up before me; how many portions he placed before me; how many cakes he offered me! All that he did, he did for my sake.' But what does the truculent guest say? 'What did I eat of his? A piece of bread, a bite of meat. What did I drink? A cup of wine. Whatever he did, he did for the sake of his wife and his children (who were sharing the meal with him).'"[201]

Given the same food, the same table, the same welcome, the same camaraderie, one guest is elated; the other slighted. One guest is full; the other unsatisfied. One guest is grateful; the other insulted. In offering this teaching, Ben Zoma seems not only to be depicting two different personalities but prodding us to ask: Which kind of guest are we? How do we receive the good in our lives? How do we react to kindnesses given to us by another? How do we respond to the hospitality of the earth?

Ben Zoma offers a simple formula for achieving happiness: be aware and savor the good. For with savoring comes gratitude; with gratitude comes enjoyment; with enjoyment comes contentment; with contentment comes peace—which is good both for us and for the earth.

Seeking inner peace through our habits of consumption is neither good for the soul nor good for the earth, for we will fail to fill the former while we denude the latter. Psychologists Amy Isham and Tim Jackson explain, "When individuals internalize the opinion that happiness and status can be achieved via the acquisition of material goods, they tend to show less environmental concern, be less inclined to engage in pro-environmental behaviours, and produce more greenhouse gases."[202]

"Judaism" Means "Thanks"

Gratitude pervades Judaism. And it is no wonder, for the very name Judah, *Yehudah*, the root of "Judaism," means "thanks." As children, some of the first words our parents teach us, some of the first words we hear, are "thank you." Upon waking up each morning, even before getting out of bed, our tradition urges us to say thank you: *modeh ani* (masculine), *modah ani* (feminine), or *odecha* (non-gendered). "Thank You, God, for restoring my soul and giving me this day."

The morning prayers that follow contain a litany of thanks for health, freedom, wisdom, strength of spirit, the constant return of the sun, the renewal of days, the intricate workings of the body. Three times a day Jews thank God for "the miracles You do for us every day, the wonders and beneficences at every moment." In special times of celebration, achievement, novelty, and more, Jews are to say the Shehecheyanu, "Thank You, God, for giving us life, sustaining us, and bringing us to this moment."

Every blessing—and Jews are to recite at least one hundred every day, throughout the day—is meant to

Mary Oliver paraphrases this sentiment this way:

Instructions for
living a life:
Pay attention.
Be astonished.
Tell about it.[203]

That, in essence, is the Shehecheyanu.

be an expression of thanks, creating in us a heightened awareness of all the gifts we receive, from a boring evening at home to a political system intact to the birth of a child.

This explains the need for such an infusion of blessings in our liturgy. Surely the rabbis did not believe God was so needy as to demand our constant displays of appreciation. Rather, tradition seems to be saying, it is *we* who need and benefit from such displays. Blessings serve as reminders to us that we are the recipients of the world's—and one another's—generosity. Being aware of the intricate orchestration that keeps the world afloat is humbling and uplifting. Such awareness can prevent us from becoming complacent, or prideful, or greedy. It can interrupt our impulse to take the world's goodness for granted or to believe that we alone are responsible for all we have achieved or that we can consume the earth's resources without concern for the consequences. Over two thousand years ago, the Torah warned, "Take care lest your heart grow haughty and you forget the Eternal your God…and you say to yourselves, 'My own power and the might of my own hand have won this wealth for me'" (Deuteronomy 8:14, 8:17).

Bachya ibn Pakuda, the twelfth-century rabbi, philosopher, and pioneering Jewish ethicist, noted this danger as well. Humans often "grow up surrounded with a superabundance of divine favors which they experience continuously and to which they become so used that they come to regard these as essential parts of their being."[204] That is, we see the flow of seasons, the confluence of bird and flower and insect emerging at the same time as automatic, perpetual, indeed so natural that we hardly think about it at all. So too regarding the worm that turns the soil, the algae and lichens that create oxygen for us to breathe, the two million red blood cells that our bodies make every second,[205] the one billion tons of rain that fall on the earth every minute, distributing its life-giving goodness.[206] Such acts become merely the invisible backdrop in which the drama of our lives is played out.

But of course they are and have always been more than that, especially now that we can affect them so profoundly. They provide the substance of our lives. Everything we eat, touch, and possess, our very bodies, bear the imprint of the workings of all parts of Creation. We know that we are not independent souls. Yet we often act as if we forget. We need the gifts of the earth and one another, both materially and spiritually. We need a village to provide for our needs, filling the gaps in our abilities and imagination. We need rivers that run clean, soils that nourish the plants, air that fills our lungs with every breath. We need trees, animals, fields, seas. Offering thanks for the wonders of the world and the presence of one another is a way to remember all that. And in remembering, we are often moved to protect it all.

Gratitude, then, is more than social or theological politeness. It drives us to act, paying life forward, creating a virtuous circle of giving, filling the world in a way monetary transactions can't. Samson Raphael Hirsch put it this way: All life exists in "one glorious chain of love, of giving and receiving.... None is by or for itself, but all things exist in continual reciprocal activities."[207] As Robin Wall Kimmerer (botanist, decorated professor, and member of the Citizen Potawatomi Nation) puts it, life is a covenant of reciprocity.[208]

Such a covenant encompasses both people and the earth, the sentient and the inanimate. Earth has brought forth so much for so long, allowing life to flourish. And we have taken so much from it for so long—sometimes unwisely. This rare world is an unearned gift whose goodness we have often poorly reciprocated. We must now tend to the needs of the earth as it has tended to ours in a grand covenant of reciprocity. As the early rabbis wrote, "Just as we receive benefits from the land, so we in turn are to be a benefit to it."[209]

God's Name El Shaddai—Divine Limits

If the earth's covenantal role is to give to us abundantly, humanity's covenantal role, and the covenantal role of each of us, is to consume modestly, gratefully, regeneratively. Our default manner of consumption should be "enough."

So important is the value of being sated with enoughness, of knowing when to stop, that the rabbis tell us "enough" is one of the attributes of God. "When Abram was ninety-nine years old, God appeared to him and said, " 'I am El Shaddai' " (Genesis 17:1). El Shaddai is an unusual name for God, appearing only seven times in the Bible. Its meaning is obscure. So the rabbis, as they are wont to do when the text is fuzzy, use their imaginations to explain.

Resh Lakish in third-century Palestine taught, "What is the meaning of 'I am El Shaddai'? It means: I am the One who said to the world "enough [*dai*]," telling the reaches of primordial Creation when to stop.[210] Life demands constraints, limits. Part of God's task in bringing forth Creation was not just in the speaking and doing but in the discipline of constraining. According to this midrash, God—the infinite in a universe of finite space and matter—saw that life needed boundaries to survive and thrive. So God placed limits on the earth, how big it should be, where the waters should go, where the land should begin, when growing should start and when it should stop. Exerting such discipline, Resh Lakish seems to suggest, is a necessary task for humanity too.

Rashi, too, explains Shaddai as meaning enough (*sha* = "that which is," and *dai* = "enough"). That is, "God provides well for all God's creatures"[211] through enoughness. God provides not excessively but fully, not extravagantly but abundantly, so that all will be sated with enough—which is, after all, all that we need.[212]

The name Shaddai, though rare in the Bible, is a constant presence in Jewish lives, chosen by tradition to be inscribed on the reverse side of every mezuzah parchment that graces the doorposts of Jewish homes. All who pass through the threshold, upon their comings and goings, pass the name of Shaddai, a marker that serves both as a blessing and as an encouragement to seek and dwell in the bounds of enoughness.

The Meanings of Manna, Spiritual Nourishment

The question remains, though, how do we know when we have had enough? There is no standard calculus that can be applied to all. The *Oxford English Dictionary* defines "enough" as "sufficient in quantity or number." But who is to say what is sufficient for me and what is sufficient for you? And how do we even know when we get there?

The Torah struggles with just this challenge as it works to transform the Israelites from a collection of slaves scarred by the experience of want into a unified collective of people trusting in the security of enoughness. It seeks to teach this through the gift of manna.

From the thirtieth day of the Exodus to the moment the Israelites entered the Land of Israel, God rained down upon them an abundance of manna, enough to feed the entire camp. Every morning (except Shabbat) the Israelites would go out from their tents and gather the manna, one omer per person, enough for each person's needs. Lest anyone become greedy and gather too much, or lest anyone misjudge and not gather enough, some divine sleight of hand would intervene to make everything right: "When they measured the manna by the omer, anyone who had gathered much had no excess, and anyone who had gathered little had no deficiency: each household had gathered as much as it needed to eat" (Exodus 16:18). Anything left overnight—that is, anything that was more than enough—would rot. Over time everyone learned their measure of enough.

What is it we can learn from this story? Perhaps that we each define our own omer, our own level of need, and that measuring enough becomes a matter of self-awareness and a measure of trust. That no one should be so gluttonous that they take more than they should, and so avoid waste and spoilage (and depriving others of what should rightfully be theirs). That no one should suffer worries and insecurity (real or imagined), causing them to feel they need to hoard today as a hedge against having enough tomorrow. That we must create and rely on systems that continually deliver and make available enough goods so that everyone can get what they need every day. While in the Torah, God was the One who oversaw all this and offered that assurance, today that job falls to us.

An Omer: The Measure of Enoughness

How much, how big, is an omer? Elsewhere in the Torah, an omer was a standard like a pound or a quart, each omer being equal to all the others. In this story, however, an omer varies, aligning its size to the needs of each person. Each individual determined the amount of their personal omer, which is to say, there is no uniformity in assessing enough. Trust and a sense of justice must rule the day. For while there cannot or should not be equality (the exact same measure of food for everyone), there should be equity (the same measure of satisfaction for all). What this story teaches is that while *enough* may be a measure of volume, *enoughness* is a measure of the soul. Manna was designed to satisfy both.

The rabbis tell us that the taste and texture of the manna changed according to the needs of the individual. Though Exodus says it tasted like wafers with honey, the rabbis tell us to the youth, it tasted like fresh bread; to the elderly, like honey wafers; to nursing babies, like milk from mothers' breasts; to the ill, like fine flour mixed with honey.

No Asceticism

Yet, enoughness should never be confused with deprivation. The Talmud takes pains to teach us that asceticism is no virtue. It chides two rabbis who flirt with, and seemingly cross over into, ascetic territory. For these rabbis, consuming something of greater value when something of lesser value is available is wasteful and wrong. "Rav Chisda said, 'When one can eat bread made of barley [a cheaper, coarser grain] but eats instead bread made of wheat [a more expensive grain], he violates *bal tashchit*, the command not to waste or destroy things.' Similarly, Rav Papa said, 'When one can drink [inexpensive] beer but drinks [more expensive] wine instead, he violates *bal tashchit*.'"[213]

But the Talmud—and modern Judaism—rush to disagree. Deprivation and abstinence are not virtues. Life offers so much for us to taste and experience. Rav, a third-century rabbi, taught, "In the world to come, we will have to account for every enjoyment offered by this world that we refused without sufficient cause."[214] Caring for the earth is indeed a necessary and uncontestable good. It is a requirement of our tradition and a desideratum of life itself. But we must not veer into the extremes of privation as we seek to avoid the harms of indulgence. Instead we are taught, "One must tend to the well-being of one's body as much as one tends to the well-being of the earth."[215] Enoughness is not about sacrifice. It is not a contest to see how little one can consume. It is a pursuit of contentment, a state in which we can say, "This is good. I am full. I have no room, no need, for anything more."[216]

Transcending Things

In the end, after all, it is not stuff that we want, but

rather what our stuff gives us and does for us, how it makes us feel. Except for the most utilitarian, necessary of purchases, we buy things to make us happy. As Csikszentmihalyi wrote, "While happiness itself is sought for its own sake, every other goal—health, beauty, money or power—is valued only because we expect that it will make us happy."[217]

We know, and research and experience confirm, that happiness does not reside in things. Dr. Laurie Santos, Yale University professor and creator of the blog *The Happiness Lab*, tells us that we often misjudge what really makes us happy and thus spend our lives chasing the wrong things. What fills us up, what brings us joy, are loving and being loved; moments of awe; a society rich in social capital that supports the "we" as much as the "me"; knowing and experiencing where and that we belong; having moments when task and self merge so we are no longer aware of time, concerns, or worries we may have; moments when we feel at one with it all.

These do not rely overly much on stuff, so both we and the earth benefit from their pursuit. We do not have to worry about having too many friends, enjoying too many sunsets, laughing too much, hiking too much. These are all, in their way, spiritual goods and, as it turns out, despite Lebow, the best and only way we can find enduring joy and enoughness in life. Viktor Frankl, the Austrian psychologist, wrote in the preface to his book *Man's Search for Meaning*, "Happiness cannot be pursued; it must ensue…as the unintended side-effect of one's personal dedication to a course greater than oneself."[218]

In a similar vein, in 1990, Csikszentmihalyi published a book titled *Flow*, thereby naming a state of bliss we all know but had not called out as a pathway to a happiness. Csikszentmihalyi defines flow as "*the state in which people are so involved in an activity that nothing else seems to matter; the experience itself is so enjoyable that people will do it even at great cost, for the sheer sake of doing it... we feel in control of our own actions, masters of our own fate... we feel a sense of exhilaration, a deep sense of enjoyment that is long cherished and becomes a landmark in memory for what life should be....* The happiest people spend much time in a state of flow."[219]

To devote oneself to work of worth, to give of one's time to a worthy cause, to lose oneself in a piece of music or the daily discipline of prayer or the practice of meditation, to immerse oneself in the intense concentration of a sport or task yields a sense of flow, a feeling of deep satisfaction and the ease of enoughness.

Shabbat

It turns out that such experiences can be found readily in the time-honored celebration of Shabbat. Every seventh day, the Sabbath, Shabbat, comes. It is a day of rest and renewal, the celebration of fullness, a day of belonging and loving and connecting to things greater than us.

Shabbat is different from the other Jewish holidays. It does not mark a historical event or seasonal harvest. It is not dependent on the phases of the moon or national celebrations. It existed before calendars came into being, before the months were given names. It is a day, the Torah tells us, woven into the fabric of time. It marks the very-good-ness, the *tov-m'od*-ness, with which Creation was endowed. It is, in short, a recurring day of enoughness.

Shabbat really is more than a day. It is an atmosphere, a way of being, a time away from time. It is the one time every week when work ceases, when the world falls away, when we live as if there is no want, as if all our needs are met, all our appetites sated. It is a day, as Abraham Joshua Heschel teaches, "where the goal is not to have but to be, not to own but to give, not to control but to share, not to subdue but to be in accord."[220] It is a day when we are fully guests in a universe that is our most gracious host.

Shabbat totally transforms space and time, creating a taste, the rabbis tell us, of the world to come. It is twenty-five hours of peace, a day of ease, a day when the primordial light of Creation shines a bit brighter on earth, when our spirits are full, our appetites sated. It is, in short, a time when we no longer need to take from the world but are simply and deeply able to witness and celebrate it. On Shabbat, we are not to farm, mine, harvest, log, fish, or otherwise disturb the earth. We are not to cook, make, manufacture, even fix things. On Shabbat, we have arrived. We rest, not out of exhaustion but because we have enough. We are full. We rest in the embrace of the earth, and the earth rests with us.

It is hard to imagine living sustainably, or creating sustainable economies, without somehow regularly experiencing the joy of enoughness. Spiritual hunger and dissatisfaction can easily lead to Lebow's vision of a good life: endless, obsessive consumption. Shabbat is an antidote to this. We don't need more, for one day a week at least, we all have all we need.

The Torah says that on the seventh day, God ceased the work of physical creation and *vayinafash* (Exodus 31:17). While this word is translated as "refreshed," it conjures up breath, the soul, the animating spirit of life (*nefesh*). In short, on Shabbat, the world is endowed with an additional soul, allowing each of us, too, to receive a *n'shamah y'teirah*, an additional soul. Perhaps it is that soul, fresh from the fullness of heaven, that enables us to feel the blessings of enoughness.

Let Us Not Be Like Noah

All of us living in this generation, here and now, hold the world's destiny in our hands. Never before has any generation been responsible for the well-being of the entire world. What we do—each of us individually and all of us collectively—matters. The time is late, the call is urgent. Nature beckons us to respond now, wholeheartedly, and not be like Noah.

> When Noah came out of the ark, he opened his eyes and saw the whole world completely destroyed. He began crying for the world and said, "God, how could you have done this?"... God replied, "Oh Noah, how different you are from the way Abraham will be. When I tell him that I plan to destroy Sodom and Gomorrah, Abraham will argue with me on their behalf.... But when I told you, Noah, that I would destroy the entire world, I lingered and delayed, so that you would speak on behalf of the world. But when you knew you would be safe in the ark, you ignored the fate of the world. Yet now you complain?" Then Noah knew that he had sinned.[221]

This midrash calls us to be like Abraham and not Noah. We are forewarned, so must speak out and protest. We must work individually and collectively on our behavior and policies, in our personal spaces, in the marketplace, in the public domain. We know enough, are capable enough, and have the technology enough to stem the worst of climate change, species extinction, soil depletion, flooding of our coastlines, degraded lands and air and water, social unrest, displaced refugees. The question is: Do we care enough? Will we work hard enough? Will we dare enough to not look away either due to a sense of futility thinking that we cannot make a difference (for we can) or by believing that we and ours will somehow be spared (for we will not).

We are called to believe in the overwhelming elegance and promise of this world, to believe that our presence here is meaningful and beneficial, that we can join our lives to something bigger than ourselves, that we can make a difference for good, and that this magnificent experiment will not falter on our watch but be handed on, in health and strength, to generations yet untold.

That is the call of *yishuv ha'olam.*

Acknowledgments

As nature thrives in a web of relations, so does a work of writing. This book has been blessed by a serendipitous interweaving of experiences.

I have been exploring the question of Judaism and sustainability for many years. My early environmental interests in the Jewish community and the interfaith field continually urged me to find the language, the prayers, the laws, and the texts that reflected the indigenous ways of Judaism and nature. Yet it wasn't until David Behrman, the publisher of Behrman House, approached me that my exploration and thoughts were fashioned into a book. David, thank you for your trust and encouragement and for inviting me to once again join the family of Behrman House authors.

It has been almost twenty years since I headed the Coalition on the Environment and Jewish Life (COEJL). That unexpected gift allowed me to learn from Jewish environmental thinkers and advocates, including Rabbi Ellen Bernstein z"l, Rabbi Eliezer Diamond, Rabbi Fred Scherlinder Dobb, Rabbi Steve Gutow, Rabbi Lawrence Troster z"l, Nigel Savage, Dr. Hava Tirosh-Samuelson, and others whose names have regrettably blurred with the passage of time.

Mirele Goldsmith, founder of Jewish Earth Alliance, whom I also met back then, became a dear friend, and continues to teach and inspire me and countless others through her inexhaustible enthusiasm, energy, and hope.

I owe a tremendous thanks to COEJL's current executive director, Rabbi Daniel Swartz, both for inspiring me with his rich early writings on Judaism and nature—some of the very first I ever read—and for his careful reading of this manuscript. This book is better for his helpful critiques and discerning eye. Any remaining mistakes of course are mine.

Thirteen years ago, I had the privilege of learning with an international cohort of people engaged in social justice and sustainability through a program called Siach. It was there I met Dr. Jeremy Benstein, a pioneer in Jewish environmentalism and a cofounder of the Heschel Center for Sustainability in Tel Aviv. He has been my teacher, my friend, and my *hevruta* (study partner) ever since. Several ideas in this book are directly attributable to my conversations with him.

I also want to offer thanks to Dr Brent Laytham, Dean of the Ecumenical Institute at St. Mary's Seminary in Baltimore, Maryland, for inviting me to teach a course on religion and sustainability; Dr. Rebecca Hancock, Associate Dean, who cotaught the course with me; and the wonderful students who continually brought enthusiasm to the class, despite the disruptive scourge of COVID.

I thank also my editor, Rabbi Deborah Bodin Cohen, for her devotion in shepherding this project from manuscript to printed book, as well as all the staff at Behrman House whom I never met. It truly takes a village to move a book from mind to shelf, and you do it with such grace and care.

And of course, I thank my precious family, my children and grandchildren, who fill me up. You know that I do not measure love in material gifts but in time and presence and devotion and laughter—and Bubbe Ema cookies (in all the necessary dietary permutations). You are my life. It is you and your future that urge me on when despair begins to darken my soul.

And my husband, who for forty-six years, has steadfastly believed in me, in my work, in my abilities, and in my writing. And who still listens even when I have nothing new to say. You are my rock.

A Daily Meditation for Preserving a Habitable World

God, everything You have made has a purpose:
air for breathing, water for drinking and bathing,
land for living and growing,
plants and animals in gracious abundance
and humanity to work and care for Your earth.
Inspire us every day to care for Your earth
so that we may fulfill Your vision,
"Not as a wasteland did God create the earth but as a home
where we may live." (Isaiah 45:18)

רִבּוֹנוֹ שֶׁל עוֹלָם, כָּל־מָה שֶׁבָּרָאתָ,
לֹא בָּרָאתָ דָּבָר אֶחָד לְבַטָּלָה.
אֲוִיר לִנְשִׁימָה, מַיִם לִשְׁתִיָּה וּלְרַחֲצָה,
אֲדָמָה לְמִחְיָה וּלְכַלְכָּלָה,
צְמָחִים וְחַיּוֹת לִרְוָחָה,
וְאָדָם "לְעָבְדָהּ וּלְשָׁמְרָהּ."
תֵּן בְּלִבֵּנוּ כָּל־יוֹם לַעֲסֹק בְּאוֹצָרְךָ בֶּאֱמוּנָה,
וְכֵן נְקַיֵּם רְצוֹנְךָ, כְּדִבְרֵי נְבִיאֲךָ:
"לֹא תֹהוּ בְרָאָהּ, לָשֶׁבֶת יְצָרָהּ." [222]

Discussion Guide

By Rabbi Daniel Swartz

Rabbi Daniel Swartz is the executive director of the Coalition on the Environment and Jewish Life (COEJL) and the spiritual leader of Temple Hesed of Scranton, Pennsylvania.

Chapter 1

1. The book starts with a parable of King David removing the stone that prevented the world from being flooded. What do you think are some of the "stones" that humans are moving which should have been left in place? What are some of the "floods" that worry you?
2. Chapter 1 discusses how our treatment of other people and our treatment of the earth are connected. How have you experienced or witnessed that connection?
3. What role, if any, do Jewish texts play in your life? What sort of wisdom might you seek from such texts?

Chapter 2

1. How would you explain *yishuv ha'olam*? Why is it important in our day and age?
2. How might your life look different if you followed the principles of *yishuv ha'olam* more closely? How might our economy look different?
3. Chapter 2 discusses the difference between a "me" and a "we" viewpoint. What are some "me" elements of modern life that you find most troubling? What would change if they took a "we" viewpoint?

Chapter 3

1. Did you know that the stories in Genesis 1 and Genesis 2 were so different? Do you personally resonate more with one or the other? Why?
2. This book makes the case that it is important that we move from a Genesis 1 "survival" viewpoint to a Genesis 2 "sustainability" viewpoint. Can you think of examples from your life when you were in "survival" mode? When you practiced sustainability? What felt different?
3. Where do you go for solace and comfort? Why do you think so many people find places in nature to be so healing?
4. How do you feel when you hear about species going extinct? What actions do you think you might take to help prevent extinctions?

Chapter 4

1. What are some natural settings that have particularly inspired you? Discuss a time when you felt a spiritual/religious connection with nature.
2. Before reading this, had you ever noticed how much the Bible focuses on the land and nature? What were some of your favorite texts or poems from this chapter?
3. If any of you have attended a Tu BiShevat seder, share the experience.
4. How do you think reciting a blessing might change or enrich how you feel about an experience in nature?
5. How might a spiritual connection with the environment lead to or strengthen activism?

Chapter 5

1. How would you explain the principle of *bal tashchit* to a friend?
2. What are some areas of your life where you feel you do a good job not wasting? What are some areas where you might improve?
3. How would our economy be different if more people acted like God was the ultimate owner of everything? How would your behavior be different?
4. What are some commandments you think should go in the list of *mitzvot bein adam l'adamah*, our behavior toward the natural world?

Chapter 6

1. What are some habits, actions, attitudes, or possessions that our society should "release" for the sake of the planet? That you personally should release?
2. Have you ever had a sabbatical or similar extended leave? If so, what were the benefits? The challenges? If not, how do you think having one might make a difference to your life?
3. One current debate about debt forgiveness in our society centers around college loans. What lessons from *sh'mitah* could be applied to this debate?
4. What might a "*sh'mitah* of the soul" ritual look like for you? What would you want to release?

Chapter 7

1. What are some ways that you show gratitude? Do you have any regular practices of gratitude? If so, how do they make you feel? If not, what sort of a practice might you now consider?
2. What are some examples of when you wanted "more" even though you had enough? What are examples of when you knew you had enough and were able to not yearn for more?
3. In line with the teachings of Ben Zoma, in what ways are you rich?
4. Given how we are draining the earth's resources, it might be tempting to try to adopt asceticism. How might that backfire?
5. This book ends with a call to action. What are some actions you want to commit to?

Notes

Prologue

1 Nell Porter Brown, "A Mind of One's Own," *Harvard Magazine*, March–April 2020, https://www.harvardmagazine.com/2020/02/h2-emily-dickinson-museum.

Chapter 1

2 Howard Schwartz, *Tree of Souls: The Mythology of Judaism* (Oxford University Press, 2004), 125.

3 Tim Searchinger, "Redirecting Agricultural Subsidies for a Sustainable Food Future," World Resources Institute, July 21, 2020, https://www.wri.org/insights/redirecting-agricultural-subsidies-sustainable-food-future.

4 Allison Winter, "Billions in Federal Farm Payments Flow to a Select Group of Producers, Report Shows," *Kansas Reflector*, February 4, 2023, https://kansasreflector.com/2023/02/04/billions-in-federal-farm-payments-flow-to-a-select-group-of-producers-report-shows/.

5 Cassandra Zimon, "Why Your Salad Costs More Than a Burger: The Truth About Government Subsidization of the Meat & Dairy Industries," New Roots Institute, January 20, 2021, https://www.newrootsinstitute.org/articles/factory-farming-subsidies.

6 Thomas Berry, "Our Way into the Future: A Communion of Subjects," in *Evening Thoughts: Reflecting on Earth as Sacred Community* (University of California Press, 2006), 17.

7 United Nations, "The 17 Goals: Sustainable Development," https://sdgs.un.org/goals.

8 Maryanne Murray Buechner, "In Ethiopia, Girls Fetch Water Instead of Going to School," *Forbes*, August 18, 2022, https://www.forbes.com/sites/unicefusa/2022/08/18/in-ethiopia-girls-fetch-water-instead-of-going-to-school Caroline Mwongera, "Empowering Women to Boost Africa's Water Security," *Africa Reflector*, August 8, 2022, https://www.un.org/africarenewal/magazine/august-2022/empowering-women-boost-africa%E2%80%99s-water-security.

9 As the Special Rapporteurs on Environment and Human Rights assert in the first two of their sixteen principles of environmental rights, governments "should ensure a safe, clean, healthy and sustainable environment in order to respect, protect and fulfill human rights, and should respect, protect and fulfill human rights in order to ensure a safe, clean, healthy and sustainable environment." David Boyd, John Knox, and Marc Limon, "#THE TIME IS NOW: The Case for Universal Recognition of the Right to a Safe, Clean, Healthy and Sustainable Environment," Universal Rights Group, February 2021, https://www.universal-rights.org/urg-policy-reports/the-time-is-now-the-case-for-universal-recognition-of-the-right-to-a-safe-clean-healthy-and-sustainable-environment/; Edith Brown Weiss and Dobbs Ferry, *In Fairness to Future Generations: International Law, Common Patrimony and Intergenerational Law* (United Nations University, 1988), 27.

10 Johan Rockström, Potsdam Institute for Climate Impact Research, https://www.pik-potsdam.de/en; Kate Raworth, Donut Hole Economics, https://www.kateraworth.com/doughnut/; Bill McKibben, http://billmckibben.com/; Herman Daly, *From Uneconomic Growth to a Steady-State Economy* (Edward Elgar Publishing, 2014); Pope Francis, *Laudato Si* (Vatican Press, 2015); Pope Francis, "Faith and Science: Towards COP26," (speech, The Vatican, April 10, 2021); Laura Sessions Stepp, "Denominations Find Common Ground in Saving the Earth," *Washington Post*, May 22, 1992, https://www.washingtonpost.com/archive/local/1992/05/23/denominations-find-common-ground-in-saving-the-earth/3ddf1620-a134-4778-8d1c-a5ec67d6c6e3/.

11 One of the most extensive efforts at guiding the public in ways to solve our climate crisis is Paul

Hawkens, "Cascade of Solutions," Project Regeneration, https://regeneration.org/solutions.

12 Dr. Charles David Keeling started recording the steady increase in atmospheric carbon dioxide in 1956, yielding the ever-upward sweep of the Keeling curve. Natural Geographic Society, "The Keeling Curve," *Natural Geographic*, October 19, 2023, https://education.nationalgeographic.org/resource/keeling-curve/. See also Philip Shabecoff, "Global Warming Has Begun, Expert Tells Senate," *New York Times*, June 24, 1988, https://www.nytimes.com/1988/06/24/us/global-warming-has-begun-expert-tells-senate.html.

13 ARCO, the Atlantic Richfield Company, created a whole Noah's Ark set, complete with Noah, thirty-six animals (eighteen pairs), an ark, and a ramp. Little could they imagine the irony of this toy—a recollection of the near destruction of the biblical world while the fossil fuels they sold were contributing to the destruction of ours. Craig Fitzgerald, "If You Were a '60s or '70s Kid, You'll Remember These Gas Station Giveaway Toys," *Bestride*, 2014, https://bestride.com/news/entertainment/gas-station-giveaway-toys.

14 UNEP, UNEP Copenhagen Climate Centre (UNEP-CCC), *Emissions Gap Report 2020* (UN Environment Programme, December 9, 2020), https://www.unep.org/interactive/emissions-gap-report/2020/.

15 Caroline Hickman et al., "Climate Anxiety in Children and Young People and Their Beliefs about Government Responses to Climate Change: A Global Survey," *Lancet Planet Health* 5 (2021): 863–73, https://www.thelancet.com/action/showPdf?pii=S2542-5196%2821%2900278-3.

16 *The Great Work: Our Way into the Future* (Bell Tower, 1999), 1.

17 Thanks to Daniel Swartz for enhancing this thought.

18 Aldo Leopold, "The Land Ethic," Aldo Leopold Foundation, https://www.aldoleopold.org/about/the-land-ethic.

Chapter 2

19 To be precise, the second day has no such affirmation, though the third day has two.

20 "Gaia Hypothesis," Wikipedia, https://courses.seas.harvard.edu/climate/eli/Courses/EPS281r/Sources/Gaia/Gaia-hypothesis-wikipedia.pdf.

21 Ellen Bernstein, *The Splendor of Creation* (Pilgrim Press, 2005), 2.

22 Lenn E. Goodman, "Respect for Nature in the Jewish Tradition," in *Judaism and Ecology*, ed. Hava Tirosh-Samuelson (Harvard University Press, 2002), 243.

23 Charles Choi, "The 10 Strangest Places Where Life Is Found on Earth," Live Science, December 3, 2010, https://www.livescience.com/29865-strangest-places-life-found.html.

24 Maimonides, *Mishneh Torah, Hilchot T'shuvah* 5:1. My thanks to Rabbi Fred Scherlinder Dobb for pointing out that Phyllis Trible once also used a similar formulation of "a part" and "apart."

25 Thomas Berry, "The Ecological Age," in *The Dream of the Earth* (Sierra Club Books, 1988), 42.

26 Rashi on *Y'vamot* 62a.

27 Yisrael Lifschitz (1782–1860) on *Pirkei Avot* 3:14.

28 Mordecai ben Avraham Yoffe, *Yoreh Deah* 249:15.

29 *Kohelet Rabbah* 7:13.

30 *Sefer Hachinuch*, Mitzvah 1.

31 Saadia Gaon, *The Book of Beliefs and Opinions*, trans. Samuel Rosenblatt (Yale University Press, 1951), 381. See also Sharon Joseph Levy, "Judaism, Population and the Environment," in *Population, Consumption and the Environment: Religious and Secular Responses*, ed. Harold G. Coward (State

University of New York Press, 1995), 81.

32 Edith Brown Weiss calls this intergenerational equity, the ethical commitment to ensure environmental equity for future generations. Louis B. Sohn and Edith Brown Weiss, "Intergenerational Equity in International Law," *American Society of International Law* 81, (April 8, 1987): 126–33, https://www.jstor.org/stable/25658355.

33 Abarbanel on Deuteronomy 22:6. So too, in a modern idiom, Hans Jonas argues in *The Imperative of Responsibility*, that humans are duty-bound to care for all Life, that which exists now and that which is yet to come. Life, writ large, he says, possesses its own imperative to be and continue being. That is the ethic that should guide our actions. See Hans Jonas, *The Imperative of Responsibility: In Search of an Ethics for the Technological Age*, trans. Hans Jonas with the collaboration of David Herr (University of Chicago Press, 1984).

34 Jonas, *Imperative of Responsibility*, 37; Hein Berdinesen, "On Hans Jonas' 'The Imperative of Responsibility,'" *Philosophia* 17 (2017): 16–28, https://philosophia-bg.com/archive/philosophia-17-2017/on-hans-jonas-the-imperative-of-responsibility/.

35 Jonas, *Imperative of Responsibility*, 11.

36 Or as it is sometimes remembered: "Never doubt that a small group of thoughtful, committed citizens can change the world. Indeed, it's the only thing that ever has." No one has been able to find the source of this quote. Indeed it may never have been uttered by Margaret Mead at all. But it is a testament to her legacy, and the optimism of the modern mind, that so many quote a version of it. See https://www.interculturalstudies.org/faq.html#quote.

37 Gur Aryeh (Yehudah Leib, the Maharal, of Prague, sixteenth century) on Deuteronomy 22:8 shows how performing one mitzvah leads to yet another mitzvah, and that mitzvah leads to yet another, so there is a cascade of goodness released into the world.

38 *Sefer Hachinuch*, 16. C. S. Lewis teaches the same thing this way: "The rule for all of us is perfectly simple. Do not waste time bothering [determining] whether you [believe you do or must] love your neighbour; [simply] act as if you did. As soon as we do this we find one of the great secrets. When you are behaving as if you loved someone, you will presently come to love him." "The Question of God: Mere Christianity (1952), Book 3, Chapters 7 & 9," PBS, https://www.pbs.org/wgbh/questionofgod/ownwords/mere2.html.

39 Jeremy Benstein, *The Way into Judaism and the Environment* (Jewish Lights, 2006), 89.

40 For more information on the successful workings of the commons, see "Elinor Ostrom's 8 Principles for Managing a Commons," *On the Commons*, 2011, https://www.onthecommons.org/magazine/elinor-ostroms-8-principles-managing-commmons/index.html.

41 "For an Economy in Service of Life," Wellbeing Economy Alliance, https://weall.org/.

42 "Robert F. Kennedy's Remarks at the University of Kansas," Wikipedia.

43 Daniel Swartz, "Israel Environment & Nature: A Brief History of Nature in Jewish Texts," Jewish Virtual Library, https://www.jewishvirtuallibrary.org/a-brief-history-of-nature-in-jewish-texts.

Chapter 3

44 I will refer to these two stories as Genesis 1 and Genesis 2.

45 Nachmanides on Leviticus 19:19, in *Commentary on the Torah by Ramban (Nachmanides)*, trans. Charles B. Chavel (Shilo, 1971–76), https://www.sefaria.org/Ramban_on_Leviticus.

46 Joseph B. Soloveitchik, *The Lonely Man of Faith* (Doubleday, 1965), 12.

47 Overpopulation is a global concern. The carrying capacity of the world is limited, after all.

Yishuv ha'olam—as noted in the previous chapter—preferences sustainability and urges this blessing of fertility to be pursued in consonance with sustainability. The November 2019 World Scientists' Warming on Climate offers the following information and guidance: "Still increasing by roughly 80 million people per year, or more than 200,000 per day, the world population must be stabilized—and, ideally, gradually reduced—within a framework that ensures social integrity. There are proven and effective policies that strengthen human rights while lowering fertility rates and lessening the impacts of population growth on GHG (greenhouse gas) emissions and biodiversity loss. These policies make family-planning services available to all people, remove barriers to their access and achieve full gender equity, including primary and secondary education as a global norm for all, especially girls and young women." William Ripple et al., "World Scientists' Warning of a Climate Emergency," *Bioscience* 70, no 1 (January 2020): 8–12. https://academic.oup.com/bioscience/article/70/1/8/5610806. See also https://www.drawdown.org/solutions/women-and-girls.

48 As told by these human authors. We can only speculate what the story would be if the animals had their say. One such fanciful tale was told sometime before the tenth century written in Arabic by members of the Islamic "Brethren of Purity," Ikhwan al-Safa, a Sufi order and subsequently translated into Hebrew by Rabbi Kalonymus ben Kalonymus, at the request of King Charles of Anjou (in France), in the year 1316. This multi-faith enterprise was then translated into English in 2005 by Rabbis Anson Laytner and Dan Bridge. Ikhwan al-Safa, *The Animals' Lawsuit Against Humanity*, trans. Rabbis Anson Laytner and Dan Bridge (Fons Vitae, 2005), https://ansonlaytner.com/the-animals-lawsuit-against-humanity/.

49 In Genesis 1, humankind were given permission only to eat plants. "God said, 'See, I give you every seed-bearing plant that is upon all the earth, and every tree that has seed-bearing fruit; they shall be yours for food. And to all the animals on land, to all the birds of the sky, and to everything that creeps on earth, in which there is the breath of life, [I give] all the green plants for food.' And it was so" (Genesis 1:29–30). Eating meat was not explicitly permitted until after the flood, in the story of Noah: "Every creature that lives shall be yours to eat; as with the green grasses, I give you all these" (Genesis 9:3).

50 Jesse Peterson, "How Animal Agriculture Is Accelerating Global Deforestation," Earth Org, February 22, 2024, https://earth.org/how-animal-agriculture-is-accelerating-global-deforestation.

51 Kristi Delynko, "What's the Beef with Water?," Denver Water, February 14, 2019, https://www.denverwater.org/tap/whats-beef-water?size=n_21_n.

52 It seems over time we have mistaken an original, justifiable meaning of exploit, to "make full use of and derive benefit from" with the subsequent, destructive meaning, to "use someone or something unfairly for one's own advantage." "Exploit," *Cambridge English Dictionary*, https://dictionary.cambridge.org/us/dictionary/english/exploit.

53 See, for example, the story of Honi, the rain-maker as well as the circle-drawer (Babylonian Talmud, *Ta'anit* 23a).

54 "Segment: Wendell Berry on His Hopes for Humanity," interview by Bill Moyers, November 29, 2013, video, 39:36, https://billmoyers.com/segment/wendell-berry-on-his-hopes-for-humanity/.

55 At least not until the Divine was provoked by the generation of Noah. Which is all the more reason why the Torah would tell the human, be careful that you care and guard the earth. For it is humans alone who can debase and threaten it. But even in the Noah story, God promises never to destroy the world again—at least not by water.

56 I thank Dr. Jeremy Benstein for sharing this insight with me.

57 A sensitive and informative reading of the distinctions between Genesis 1 and 2 can be found in Theodore Hiebert, *The Yahwist's Landscape* (Oxford University Press, 1996).

58 For a similar argument of needing to move from our value framework to another, see Jonas Salk's discussion of the S-curve in Roman Krznaric's *The Good Ancestor: A Radical Prescription for Long-Term Thinking.* Garrett Hardin (in "The Tragedy of the Commons") wrote that "the morality of an act is a function of the state of the system at the time it is performed," to which we respond: yes and no. On the one hand, Judaism teaches it is never okay to wantonly destroy things—whether natural or human-made, whether the item occurs in abundance or is the last one of its kind. For in part, Judaism is as much concerned about the intent and mindfulness of the destroyer as about the item being destroyed. And it is never acceptable for a willful, wasteful spirit to wantonly destroy anything. (We will explore this more in chapter 5.) On the other hand, the context in which decisions are made is critical. Fighting COVID-19, for example, has created an overabundance of materials that turn into waste (personal protective equipment, masks, disposal syringes for vaccinations, test kits). Such abundant waste would otherwise be unacceptable in a world struggling to consume less and recycle more. But who would say that such consumption is inappropriate when it is in service of saving millions of lives? Roman Krznaric, *The Good Ancestor: A Radical Prescription for Long-Term Thinking* (The Experiment, 2020), 122–23; Garrett Hardin, "Tragedy of the Commons," *Science* 162, no. 3859 (December 13, 1968): 1243–48.

59 Rabbi Joseph ibn Kaspi, in his Bible commentary, *Matzref La-Kesef* on Deuteronomy 22:6–7. Thanks to Rabbi Daniel Swartz for bringing this text to my attention.

60 Mary Oliver, *Winter Hours, Upstream: Selected Essays* (Penguin Press, 2016), 154.

61 Rabbi Ben-Zion Hai Ouziel, the Sephardic chief rabbi of Israel, said it this way: "*Yishuvo shel olam* is a precondition and prerequisite to achieving the moral footing of our lives."

Chapter 4

62 Manfred Gerstenfeld and Netanel Lederberg, "Nature and the Environment in Hasidic Sources," *Jewish Environmental Perspectives* (October 2002): 54, https://www.jcpa.org/art/jep5.html, quoting the Baal Shem Tov, *Zava'at ha-Rivash*, photocopy (Brooklyn: Otsar Hahasidim, 1991), section 141, p. 54 (Hebrew). Some scholars believe that many of the sayings in this book are probably not attributable to the Baal Shem Tov but to Rabbi Dov Ber, the Maggid of Mezeritch, his main disciple. See also David Mevorach Seidenberg, "Building the Body of the Shekhinah: Re-enchantment and Redemption of the Natural World in Hasidic Thought," in *A New Hasidism: Branches*, ed. Arthur Green and Ariel Evan Mayse (Philadelphia: Jewish Publication Society, 2019).

63 "Divine," *Merriam-Webster Dictionary*, https://www.merriam-webster.com/dictionary/divine.

64 "To bless the new moon," the Talmud tells us, "at the proper time is like greeting the Divine Presence" (Babylonian Talmud, *Sanhedrin* 42a).

65 *B'reishit Rabbah* 10:6.

66 This lovely phrase is from Julia Watts Belser, *Power, Ethics and Ecology in Jewish Late Antiquity* (Cambridge University Press, 2015), 40.

67 *Pirkei Avot* 3:9 or in some counts 3:7.

68 See Jeremy Benstein's excellent article on this. Jeremy Benstein, "'One, Walking and Studying…' Nature vs. Torah," in *Judaism and Environmental Ethics: A Reader* (Lexington Books, 2001), 206–29.

69 Benstein, "'One, Walking and Studying,'" 215.

70 "How Chernobyl Has Become an Unexpected Haven for Wildlife," UN Environment Programme, September 16, 2020, https://www.unep.org/news-and-stories/story/how-chernobyl-has-become-unexpected-haven-wildlife.

71 Babylonian Talmud, *M'nachot* 29b.

72 Daniel C. Matt, *The Essential Kabbalah: The Heart of Jewish Mysticism* (Castle Books, 1997), 90.

73 Abraham Joshua Heschel, *God in Search of Man* (Farrar, Straus and Giroux, 1976), 105–6.

74 Heschel, *God in Search of Man*, 106.

75 "When we succeed in slowing down our experience enough to pay attention to our individual senses, such that our feelings and their interaction with the whole of the world become perceptible, something astounding happens: With a sigh of tender gratitude, we understand we have always already been the recipients of a gift." Andreas Weber, *Matter and Desire: An Erotic Ecology* (Chelsea Green Publishing, 2017), 192.

76 Soloveitchik, *The Lonely Man of Faith*, 23.

77 Bahya ben Yosef ibn Pakuda, *Duties of the Heart*, trans. Moses Hyamson, (Boys Town Publishers, 1985), 137, 139.

78 "Gezer Calendar," Wikipedia, https://en.wikipedia.org/wiki/Gezer_calendar.

79 Referring to sexual prohibitions in these verses.

80 Johannes Pederson, quoted in Everett Gendler, "A Sentient Universe," in *Ecology and the Jewish Spirit: Where Nature and the Sacred Meet*, ed. Ellen Bernstein (Jewish Lights, 1998), 60.

81 Liberty Hyde Bailey, *The Holy Earth: The Birth of a New Land Ethic* (Counterpoint, 2015), 4.

82 Herman Daly, *Beyond Growth* (Beacon Press, 1996), 221.

83 Summer Allen, *The Science of Awe* (John Templeton Foundation, September 2018), https://www.templeton.org/wp-content/uploads/2018/08/White-Paper_Awe_FINAL.pdf.

84 Interestingly, the Julian calendar also has a twenty-eight-year cycle in which the assignment of days of the weeks on the numbers of the months repeats in its entirety.

85 Eugene Borowitz, quoted in Barbara Lerman Golomb, "Reconnecting with Nature to Sustain Ourselves," in *The Mountains Shall Drip Wine: Jews and the Environment*, ed. Leonard J. Greenspoon (Creighton University Press, 2009), 66.

86 Daniel Swartz, "Israel Environment & Nature: A Brief History of Nature in Jewish Texts," Jewish Virtual Library, https://www.jewishvirtuallibrary.org/a-brief-history-of-nature-in-jewish-texts.

87 G. K. Chesterton, "The Ethics of Elfland," in *Orthodoxy* (Project Gutenberg), 98, https://www.gutenberg.org/ebooks/130.

88 Perhaps as many as 180 verses. *A. O. Scott, "'Hallelujah' Review: From Leonard Cohen to Cale to Buckley to Shrek," New York Times,* June 30, 2022, https://www.nytimes.com/2022/06/30/movies/hallelujah-leonard-cohen-a-journey-a-song-review.html.

89 Leah Goldberg, "The Poems of the End of the Journey," in *Poems II* (Hakibbutz Hame'uchad, 1986), 154.

90 *Carl Sagan's Cosmos*, "We Are a Way for the Cosmos to Know Itself," June 6, 2018, YouTube video, https://www.youtube.com/watch?v=rWnA4XLrMWA.

91 References to some form of *Perek Shirah* may be found as early as the tenth century. Printed commentaries go back as far as the sixteenth century. Wilhem Bacher and Judah David Eisenstein, "Shirah, Perek (Pirke)," *Jewish Encyclopedia*, https://www.jewishencyclopedia.com/articles/13588-shirah-perek-pirke.

92 *Perek Shirah* 1.

93 *Perek Shirah* 3.

94 John Burroughs, *The Art of Seeing Things* (Syracuse University Press, 2001), 9–10.

95 Hava Tirosh Samuelson, "Religion and Ecology: Can the Climate Change?," *Daedalus* 130, no. 4 (Fall 2001): 115.

96 *P'ri Eitz Hadar* 1.

97 "Rabbi Nachman of Bratslav, Prayer for Nature," Sefaria, https://www.sefaria.org/sheets/114332?lang=bi. See also Reb Nosson of Breslov, *The Flame of the Heart: Prayers of a Chasidic Mystic Reb Nosson of Breslov*, trans. David Sears (Jewish Lights, 2006), 64.

98 This saying is quoted in the name of Rav Kook. R. Shear Yashuv Cohen, "The Cutting of Trees in Times of Peace and War," *Tehumim* 4 (1983): 45.

99 Ben Zion Bokser, *The Lights of Penitence, the Moral Principles, Lights of Holiness, Essays, Letters, and Poems*, trans. Abraham Isaac Kook, (Paulist Press, 1978), 26.

100 A. D. Gordon, quoted in Reuben Wallenrod, "The Teachings of A. D. Gordon (1856–1922)," *Jewish Social Studies 7*, no. 4 (October 1945): 337–56.

101 *Anthology of Modern Hebrew Poetry*, trans. A. C Jacobs (Institute for the Translation of Hebrew Literature and Israel Universities Press, 1966).

102 Saul Tchernikovsky, quoted in Swartz, "Israel Environment & Nature."

103 It finds that if techniques to improve soil carbon were rolled out at the maximum assumed level worldwide, they could remove up to 5.5bn tonnes of CO2e a year. Daisy Dunne, "Restoring Soils Could Remove up to '5.5bn Tonnes' of Greenhouse Gases Every Year," Carbon Brief, March 16, 2020, https://www.carbonbrief.org/restoring-soils-could-remove-up-to-5-5bn-tonnes-of-greenhouse-gases-every-year; Jessica Colarossi, "City Trees and Soil Are Sucking More Carbon Out of the Atmosphere Than Previously Thought," *The Brink*, February 16, 2022, https://www.bu.edu/articles/2022/city-trees-and-soil-are-sucking-more-carbon-out-of-the-atmosphere-than-previously-thought.

104 Moshe ben Yehudah Ibn Machir, *Seder Hayom* (1599), commentary to *Mishnah Avot* 5:21.

Chapter 5

105 Sir William Blackstone, the famed eighteenth-century English jurist and author of the influential *Commentaries on the Laws of England* (1765), describes a seemingly irrefutable, fundamental, attribute of the human spirit: "There is nothing which so generally strikes the imagination, and engages the affections of mankind, as the right of property; or that sole and despotic dominion which one man claims and exercises over the external things of the world, in total exclusion of the right of any other individual in the universe."

106 Hannah Arendt, quoted in Mihaly Csikszentmihalyi and Eugene Rochberg-Halton, *The Meaning of Things: Domestic Symbols and the Self* (Cambridge University Press, 1999), 16. Even more, Csikszentmihalyi writes, "Not surprisingly, the clothes one wears, the car one drives and the furnishings of one's home, are all expressions of one's self, even when they act as disguises rather than as reflections" (14–15).

107 Babylonian Talmud, *Bava M'tzia*. The Torah offers a universal rule of returning lost objects in Deuteronomy 22:1–3.

108 "Despotic dominion" is a phrase used by the eighteenth-century English jurist in his landmark commentary on the laws of England (Blackstone, *Commentaries on the Laws of England*). We will discuss this a bit more in chapter 6.

109 "A usufruct is a legal right accorded to a person or party that confers the temporary right to use and derive income or benefit from someone else's property.... While the usufructuary has the right to use the property, they cannot damage or destroy it or dispose of the property."

Will Kenton, "What Is Usufruct? How It Works with Property Use and Example," Investopedia, April 2, 2021. https://www.investopedia.com/terms/u/usufruct.asp.

110 One could argue that the sidewalk is part of the commons and that therefore the pedestrian is part owner of that space along with everyone else who pays taxes and shares in the commons. But even if so, ownership in the commons confers different rights than private ownership. And so to sidestep this argument for the sake of this chapter, let's assume the pedestrian is a tourist who has no ownership stake in this commons.

111 Loren Eiseley, quoted in Herman Daly, *Beyond Growth* (Beacon Press, 1996), 213.

112 These last words possess a double-entendre, meaning both: all is created in accord with God's word, and all is conducted in accord with God's will.

113 "Though the earth, and all inferior creatures, be common to all men, yet every man has a property in his own person: this no body has any right to but himself. The labour of his body, and the work of his hands, we may say, are properly his. Whatsoever then he removes out of the state that nature hath provided, and left it in, he hath mixed his labour with, and joined to it something that is his own, and thereby makes it his property." John Locke, *Second Treatise on Government* (1689).

114 The Torah here means food for human consumption. Yet we know now that all trees serve as "food" of some sort for other creatures, animate and not, including through their mycorrhizal network. Today's expanded understanding of the value of trees to their surrounding ecosystems may help us extend the application of this verse.

115 Locke, *Second Treatise on Government*, chapter 5.

116 Locke, *Second Treatise on Government*, chapter 2.

117 "The reason for this entire legislation," the *Tur Ha'aroch* says, "and specifically for its being mentioned at this point, is that in war soldiers are in the habit of wreaking havoc all over, without regard to the ecological damage they cause by doing so. The Torah, therefore, goes on record that even in war, one must be concerned with what will be needed after the war has ended. *Tur Ha'aroch*, Deuteronomy 20:19.

118 Samson Raphael Hirsch, *The Pentateuch*, (Judaica Press, 1973), Deuteronomy 20:20.

119 Clare Miflin, "If Nature Doesn't Need Trash, Neither Do We," Grist, October 24, 2018, https://grist.org/article/if-nature-doesnt-need-trash-neither-do-we/.

120 "Waste," *Oxford Online Dictionary*.

121 Samson Raphael Hirsch, *Horeb*, vol. 2 (Soncino Press, 1962), 279.

122 This is one reason why cutting a ribbon or garment as an act of mourning is so powerful. By not only permitting but encouraging the marring of a garment after the death of a loved one, something that would otherwise be a clear violation of this fundamental prohibition, the tradition is showing us that it realizes the rupturing power of death and how we temporarily live in a different reality in the wake of death.

123 Hirsch, *Horeb*, vol. 2, 279. See also Samson Raphael Hirsch, *The Pentateuch*, vol. 5 (Judaica Press, 1973), 395.

124 In 1987, the United Nations Brundtland Commission defined sustainability as "meeting the needs of the present without compromising the ability of future generations to meet their own needs." "Sustainability," United Nations Academic Impact, https://www.un.org/en/academic-impact/sustainability. See also Edith Brown Weiss, "In Fairness to Future Generations and Sustainable Development," *American University International Law Review* 8, no. 1 (1992): 19–26.

125 Samson Raphael Hirsch, *The Nineteen Letters on Judaism*, trans. Bernard Drachman, (Feldheim, 1969), 36.

126 Maimonides, *Mishneh Torah, Hilchot M'lachim* 6:8–10.

127 Hirsch, *Horeb*, vol. 2, 280.

128 One of the premier proponents of intergenerational equity today is the legal scholar Edith Brown Weiss. She has written extensively on climate change, intergenerational equity, and international law. Several countries, including Israel, have experimented with creating governmental ministries for future generations to guarantee that the interests of future generations are represented in decisions that are made today. Wales has been the most successful in integrating this position into its decision-making process.

129 "Maastricht Principles on the Human Rights of Future Generations," Maastricht University, https://www.maastrichtuniversity.nl/research/maastricht-centre-human-rights/maastricht-principles-and-guidelines; "Futures Tools for Intergenerational Equity," UNESCO, https://foresight.unglobalpulse.net/intergenerational-equity/.

130 Maimonides, *Mishneh Torah, Hilchot M'lachim* 6:8.

131 "Externality," *Merriam-Webster Dictionary*, https://www.merriam-webster.com/dictionary/externality.

132 Daly, *Beyond Growth*, 222.

133 Suzanne Simard, *Finding the Mother Tree* (Knopf, 2021).

134 *Pirkei D'Rabbi Eliezer*, chapters 33–34.

135 Andreas Weber, *Matter and Desire: An Erotic Ecology*, (Chelsea Green Publishing, 2017), 35.

136 *Sefer Hachinuch*, 529.

137 Such as wasting water while brushing one's teeth or taking excessively long showers to injecting millions of gallons of fresh water into fracking wells, creating more packaging than necessary, using single-use disposables when reusable items are just as good, or designing products that cannot be readily repaired.

138 Such as destroying mountains, valleys, and waterways to get at the minerals within.

139 Such as clear-cutting the Amazon to create grazing areas for cattle when it is of greater utility as a forest than pasture, and reclaiming spent pastureland for grazing.

140 That is, any unwise return on the investment.

141 Such as hoarding or otherwise preventing wise use.

142 *Cattle, Cleared Forests, and Climate Change* (Union of Concerned Scientists, 2016), https://www.ucsusa.org/sites/default/files/attach/2016/09/ucs-cattle-cleared-forests-climate-change-2016.pdf.

143 Theodore Roosevelt, speech at Grand Canyon, May 6, 1903, in *American Earth: Environmental Writing since Thoreau*, ed. Bill McKibben (Literary Classics of the US, 2008), 133.

144 Joseph Soloveitchik, *Halakhic Man*, trans. Lawrence Kaplan (Jewish Publication Society of America, 1983), 46–47, https://archive.org/details/halakhicman00solo/page/46/mode/2up.

145 "The 17 Goals: Sustainable Development," United Nations, https://sdgs.un.org/goals.

146 "Principles of Environmental Justice," First National People of Color Environmental Leadership Summit (Washington DC, October 24–27, 1991), https://www.ejnet.org/ej/principles.html.

147 Given the Jewish propensity for eighteen, *chai*, "life," once we get to seventeen, why not push for eighteen?

148 Leviticus 25:23, "You are but strangers resident with Me."

149 Leviticus 25:23, "The land is Mine."

150 By domain I mean any distinct area of earth, be it land, water, or air, that one's act will be benefiting from and impacting.

151 The Talmud speaks of such waste when it records Rabbi Zutra's teaching: one who covers an oil

lamp or uncovers a naphtha lamp violates *bal tashchit* (causing the wicks to burn too fast) (Babylonian Talmud, *Shabbat* 67b).

152 One mantra is to abide by the nine "R's": Refuse, reduce, reuse, rethink, repair, refurbish, repurpose, recycle, recover.

153 See also the UN World Charter for Nature, which speaks in similar terms of this call to live within the bounds of nature's needs. "World Charter for Nature," UN General Assembly (October 28, 1982), https://ejcj.orfaleacenter.ucsb.edu/wp-content/uploads/2018/03/1982.-UN-World-Charter-for-Nature-1982.pdf.

Chapter 6

154 Dr. Jeremy Benstein, "Alma D'atei: The World That Is Coming; Reflections on Power, Knowledge, Wisdom and Progress," in *The Good in Nature and Humanity: Connecting Science, Religion, and Spirituality with the Natural World*, ed. Stephen R. Kellert and Timothy J. Farnham (Island Press, 2002).

155 Leviticus 19:9–10, 23:22; Deuteronomy 24:19. *Leket* and *shich'chah* indicate the farmer has passed the threshold "enough," else neither would exist. Stalks are dropped when we try to stuff too much in our arms. Bundled sheaves might be forgotten only when we have enough in our storehouses already. Otherwise, we would not have the luxury of forgetting desperately needed food.

156 Gerald Jacob Blidstein, "Man and Nature in the Sabbatical Year," *Tradition* 9, no. 4 (1966): 50.

157 The Talmud teaches, "The Torah speaks in the language of humankind" (Babylonian Talmud, *B'rachot* 31b, which generally is taken to mean that it speaks in the idiom of its original audience and thus demands that future generations not be literalists. We should not always be bound by the particular details but by the enduring meaning and message behind the words. And we should "translate" that meaning into contemporary idioms. I take the saying to also mean that the Torah is savvy to the limits of the human spirit—and while it seeks to push that spirit as far as it can go, it knows it will lose its authority if it radically oversteps them.

158 "Why Regenerative Agriculture?," Regeneration International, https://regenerationinternational.org/why-regenerative-agriculture/.

159 "Why Regenerative Agriculture?"

160 "The History of Lending Discrimination," Investopedia, https://www.investopedia.com/the-history-of-lending-discrimination-5076948; Interfaith Rainforest Initiative, https://www.interfaithrainforest.org.

161 Yet the creation of a permanent Israelite underclass due to debt was unacceptable to the biblical authors. So one's debt service could never last for more than seven years: "When you acquire an Israelite debt servant, that person shall serve six years—and shall go free in the seventh year" (Exodus 21:2). It seems by this formulation that this seventh year was not the *sh'mitah* year but calculated based on the number of servant years.

162 The Torah promises that God will answer the recalcitrant in biblical times. But when the early rabbis witnessed the needy being unable to secure the loans they needed and the threat of divine retribution to the would-be lender was not effective, they determined they needed to act. Hillel (who lived in the first century CE) instituted the *prozbul*, a procedure that transferred the rights of debt collection to the court, thereby circumventing the command of the Torah that addressed the individual lender. The court, not the lender, could demand repayment during or after the *sh'mitah* year. Upon the court receiving the repayment, the monies would be returned to the lender. This neatly encouraged the lenders to abide by the biblical law of generously lending to needy members of the tribe by protecting

them from financial loss. Of course, once the *prozbul* was in place, the borrower might no longer be protected from accumulated debt. Was favoring the interests of the lender the best way to protect the interests of the borrower? One could challenge that. But then, not having access to money to borrow is no win either. Without the hand of God to entice the lenders to lend, what would you do?

Debt forgiveness has strong advocates today as well. The burden of medical debt, educational debt, debt incurred through shady lending practices, the national debt of third world countries, debt incurred because of a failed business endeavor or an ill-timed investment, is increasingly being seen today as a matter of structural inequity. Intergenerational wealth transfer, when grandparents and parents are able to leave significant financial resources to their children and grandchildren, is a blessing for those fortunate enough to be the recipients. But what about those who are not so lucky? How are we to assist them in their pursuit of life?

163 Melanie Hanson, "Effects of Cancelling Student Loan Debt," Education Data Initiative, March 31, 2023, https://educationdata.org/what-happens-if-student-loan-debt-is-canceled.

164 National University of Singapore, "Debt Relief Improves Psychological and Cognitive Function, Enabling Better Decision-Making," Science*Daily*, March 26, 2019, https://www.sciencedaily.com/releases/2019/03/190326105700.htm.

165 Erica Hogan, "Why Debt Relief Matters to the Wealthy West," Carnegie Endowment for International Peace, January 17, 2024, https://carnegieendowment.org/research/2024/01/why-debt-relief-matters-to-the-wealthy-west?lang=en.

166 I thank my friend Emily Gaines Demsky for this insight.

167 See https://zenhabits.net/simplespirit/ for some benefits of decluttering. As Mihaly Csikszentmihalyi and Eugene Rochberg-Halton write, "Household objects are chosen and could be freely discarded if they produced too much conflict with the self. Thus household objects constitute an ecology of signs that reflects as well as *shapes* the pattern of the owner's self." Leo Babauta, "Simplicity as Spiritual Practice: Declutter for Deep Personal Growth," Zen Habits, https://zenhabits.net/simplespirit/; Mihaly Csikszentmihalyi and Eugene Rochberg-Halton, *The Meaning of Things: Domestic Symbols and the Self* (Cambridge University Press, 1999), 17.

168 Emilie Le Beau Lucchesi, "The Unbearable Heaviness of Clutter," *New York Times*, January 3, 2019, https://www.nytimes.com/2019/01/03/well/mind/clutter-stress-procrastination-psychology.html.

169 No twenty-first-century discussion on stuff would be complete without acknowledging the home organizing phenomenon of Maria Kondo. Kondo acknowledges that with three children and a small empire to run she has little time for organizing. What gives her pleasure is time well spent with her family. Still, she believes a well-ordered exterior contributes to a happier spirit. But if a neat home is not within reach, then she offers other ways to find calm and happiness. Jura Koncius, "Marie Kondo's Life Is Messier Now—and She's Fine with It," *Washington Post*, January 26, 2023, https://www.washingtonpost.com/home/2023/01/26/marie-kondo-kurashi-inner-calm/.

170 Abraham Isaac Kook, *Shabbat Ha'aretz*, trans. Yedidyah Sinclair, (Koren, 2022), 90.

171 To learn more about *sh'mitah* and ways it can be practiced today, see *The Shmita Sourcebook*, 3rd ed. (Hazon, 2021), https://hazon.org/resource/updated-shmita-sourcebook/.

Chapter 7

172 Victor Lebow, "Price Competition in 1955," *Journal of Retailing* (Spring 1955), http://www.ablemesh.co.uk/PDFs/journal-of-retailing1955.pdf.

173 Jean Baudrillard, *The Consumer Society: Myths and Structures* (SAGE, 1998), 208.

174 "We Bring Good Things to Life" was a slogan used by General Electric between 1979 and 2003.

175 Charles David Keeling, "The Concentration and Isotopic Abundances of Carbon Dioxide in the Atmosphere," *Tellus* 12, no. 2 (June 1960).

176 American Chemical Society National Historic Chemical Landmarks, "The Keeling Curve: Carbon Dioxide Measurements at Mauna Loa," ACS, https://www.acs.org/education/whatischemistry/landmarks/keeling-curve.html.

177 Kate Raworth, "A Healthy Economy Should Be Designed to Thrive, Not Grow," TED video, June 4, 2018, https://www.youtube.com/watch?v=Rhcrbcg8HBw; Paul Hawken, *The Ecology of Commerce* (Harper Collins, 1993), 13.

178 Gaylord Nelson, Susan M. Campbell, and Paul A. Wozniak, *Beyond Earth Day: Fulfilling the Promise* (University of Wisconsin Press, 2002), 18.

179 Will Sullivan, "Humans Have Shifted Earth's Axis by Pumping Lots of Groundwater," *Smithsonian Magazine*, June 22, 2023, https://www.smithsonianmag.com/smart-news/humans-have-shifted-earths-axis-by-pumping-lots-of-groundwater-180982403.

180 "Planetary Boundaries," Stockholm Resilience Center, https://www.stockholmresilience.org/research/planetary-boundaries.html.

181 "Earth Overshoot Day is computed by dividing the planet's biocapacity (the amount of ecological resources Earth is able to generate that year), by humanity's Ecological Footprint (humanity's demand for that year), and multiplying by 365, the number of days in a year." David Lin, Leopold Wambersie, and Mathis Wackernagel, "Estimating the Date of Earth Overshoot Day 2023," Global Footprint Network, May 2023, https://www.overshootday.org/content/uploads/2023/06/Earth-Overshoot-Day-2023-Nowcast-Report.pdf.

182 "Past Earth Overshoot Days," Earth Overshoot Day, https://www.overshootday.org/newsroom/past-earth-overshoot-days/.

183 Joseph ben Abraham Gikatilla (thirteenth century), *Sefer Hameshalim*, #128.

184 Isabella Gerretsen, "Microplastics Are Everywhere: Is It Possible to Reduce Our Exposure?," BBC, January 10, 2024, https://www.bbc.com/future/article/20240110-microplastics-are-everywhere-is-it-possible-to-reduce-our-exposure.

185 Damian Carrington, "Microplastics Found in Human Blood for First Time," *The Guardian*, March 24, 2022, https://www.theguardian.com/environment/2022/mar/24/microplastics-found-in-human-blood-for-first-time.

186 Damian Carrington, "Microplastics Found in Every Human Placenta Tested in Study," *The Guardian*, February 27, 2024, https://www.theguardian.com/environment/2024/feb/27/microplastics-found-every-human-placenta-tested-study-health-impact.

187 Theresa Machemer, "Humans Have Altered 97 Percent of Earth's Land Through Habitat and Species Loss," *Smithsonian Magazine*, April 20, 2021, https://www.smithsonianmag.com/smart-news/humans-have-altered-97-percent-earths-land-through-habitat-and-species-loss-180977542.

188 But not equitably. In America, members of lower-income communities and communities of color tend to live in neighborhoods with greater environmental degradation caused by their—undesired—proximity to sources of pollution, the detritus from the infrastructure that is designed to serve us all. The accepted calculus that we may poison some of us (who live in a place others consider "away") for the benefit all is a devil's bargain. It is both unjust and erroneous, for in the end, we will all suffer.

Even more, those who least harm the climate are the ones who suffer environmental degradation the most. "Households with incomes in the top 10 percent generate roughly 36 to 45 percent of global emissions, while households with incomes in the bottom 50 percent contribute just 13 to 15 percent."

(Sujata Gupta, "Eating meat is the Western norm. But norms can change," *Science News*, May 11, 2022. https://www.sciencenews.org/article/eating-meat-diet-western-norm-change-food-preferences) These, the most affected, are also the least able to protect themselves from the impact of environmental degradation and climate change. That is the opposite of the Torah's teachings about sharing the bounty of the fields. Owner and gleaner both eat from the same land and gather in the same harvest.

189 Babylonian Talmud, *Bava Kama* 50b.

190 Robert Waldinger, "Author Talks: The World's Longest Study of Adult Development Finds the Key to Happy Living," interview by Molly Liebergall, McKinsey & Company, https://www.mckinsey.com/featured-insights/mckinsey-on-books/author-talks-the-worlds-longest-study-of-adult-development-finds-the-key-to-happy-living. See also Marc S. Schulz and Robert J. Waldinger, *The Good Life: Lessons from the World's Longest Scientific Study of Happiness* (Simon & Schuster, 2023).

191 Amy Isham and Tim Jackson, "Finding Flow: Exploring the Potential for Sustainable Fulfillment," *Lancet Planet Health* 6 (January 2022): e66.

192 Samson Raphael Hirsch, *Horeb,* vol. 1 (Soncino Press, 1962), 44. The writer of Ecclesiastes also reminds us, "A lover of money never has his fill of money, nor a lover of wealth his fill of income. That too is futile" (5:9).

193 Jeanna Smiley, "Can Retail Therapy Actually Be Helpful?," Very Well Health, February 25, 2022, https://www.verywellhealth.com/retail-therapy-5217208.

194 Pope Francis, *Laudato Si*, para. 217 and para. 204. He puts this thought more bluntly when he writes, "The emptier a person's heart is, the more he or she needs things to buy, own and consume."

195 The Chilean economist Manfred Max-Neef urges us to distinguish between needs—finite and universal and satisfiers—many and culturally determined. He argues that understanding the difference, distinguishing what satisfiers are appropriate to feed distinct needs can help us build a more sustainable world.

196 "It is important to note that the personal resources accrued during states of positive emotions are durable—they outlast the transient emotional states that led to their acquisition." Robert A. Emmons and Michael E. McCullough, *The Psychology of Gratitude* (Oxford University Press, 2004), 149.

197 Mihaly Csikszentmihalyi , *Flow: The Psychology of Optimal Experience* (Harper Perennial, 1990).

198 "G.K. Chesterton: Quotes," Goodreads, https://www.goodreads.com/quotes/122391-here-dies-another-day-during-which-i-have-had-eyes.

199 *Tosefta B'rachot* 6:5.

200 *Pirkei Avot* 4:1.

201 *Tosefta B'rachot* 6.

202 Isham and Jackson, "Finding Flow," 66–74.

203 Mary Oliver, *Red Bird: Poems* (Beacon Press, 2008), 37.

204 Bachya ben Asher, *Rabbeinu Bahya*, trans. Eliyahu Munk (Urim, 1998), *B'reishit* 17:1.

205 "How Your Body Replaces Blood," NHS, https://www.blood.co.uk/the-donation-process/after-your-donation/how-your-body-replaces-blood.

206 Charlotte Smith, "Ten Stunning Facts about Rain to Pour Over," The Weather Channel, November 22, 2017, https://weather.com/en-GB/unitedkingdom/weather/news/ten-stunning-facts-about-rain-to-pour-over-surprising-interesting-weather.

207 Samson Raphael Hirsch, *The Nineteen Letters of Ben Uziel* (Kessinger, 2007), 36.

208 See, for example, "The Covenant of Reciprocity," in *The Wiley Blackwell Companion to Religion and Ecology*, ed. John Hart (John Wiley, 2017), 368–81.

209 Jerusalem Talmud, *Sh'vi'it* 9:1.

210 Babylonian Talmud, *Chagigah* 12a.

211 Bachya ben Asher, *Rabbeinu Bahya, B'reishit* 17:1.

212 This match between the world's giving of abundance and our consuming just enough is found in the intimate blessing Jacob gave his son Joseph: "May Shaddai bless you with the blessings from above and the blessings from the deep, blessings of the breast [*shadayim*] and blessings of the womb" (Genesis 49:25). Given all the other names of God the author could have chosen, selecting "Shaddai" seems intentionally evocative. It conjures up the image of God as a generous, tender mother of a nursing child. And as many nursing mothers know, the breast can produce milk in abundance, though the child drinks only enough—neither seeking nor consuming more than they need. So, the blessing seems to be saying, is the way the world is to us and we should be to the world.

213 Rav Chisda seems to have been no stranger to the specter of poverty. "Rav Chisda said: A student of a Torah academy who does not have much bread should not cut it into thin slices; rather, he should eat what he has in one helping. Rav Chisda said: A student of a Torah academy who does not have much bread should not break it for guests. What is the reason? As he will not do so in a generous manner. Rav Chisda said: Originally, I would not break bread until I placed my hand in the entire dish to assure that I found that there was enough bread to meet my needs" (Babylonian Talmud, *Shabbat* 140b:8–9).

214 Jerusalem Talmud, *Kiddushin* 4:12.

215 Babylonian Talmud, *Shabbat* 129a.

216 The greatest Jewish liturgical paean to enoughness has to be the Passover song Dayeinu—"Enough, we are good." It reviews a litany of kindnesses that God did for the Jewish people in their trek from Egypt to the Promised Land. And after each step, each kindness, there is a hearty refrain of *Dayeinu*—"Enough! All good!" There are fifteen stanzas (corresponding to the fifteen steps of ascent that led to the Temple in Jerusalem?) recounting all the ways God guided and protected the Israelites on their journey: "If you had only brought of us out Egypt… if you had not split the Red Sea… if you had not given us the manna… if you had not given us the Torah."

But of course, it would not have been enough if any of these things had not happened. If any of those had failed, the Jewish enterprise would have been lost. So the question is, what are we to learn from Dayeinu? Solomon Schimmel offers this explanation: "We should be grateful for benevolent efforts expended on our behalf even if they did not ultimately come to fruition. The Israelites would have owed God gratitude for each effort or action He took on their behalf even if these actions would not have reached all of their ultimate goals." Robert Emmons and Michael E. McCullough, "Gratitude in Judaism," in *The Psychology of Gratitude* (Oxford University Press, 2004), 41.

217 Csikszentmihalyi, *Flow,* 16.

218 Viktor Frankl, *Man's Search for Meaning* (Touchstone: Simon & Schuster, 1984), 12.

219 Csikszentmihalyi, *Flow,* 4.

220 Abraham Joshua Heschel, *The Sabbath: Its Meaning for Modern Man* (Shambhala, 2003), ix.

221 *Zohar Chadash, Noach*: 109–12.

Daily Meditation

222 By Rabbi Avram Reisner and Rabbi Nina Beth Cardin, Courtesy of the Rabbinical Assembly, Lev Shalem for Weekdays, 2025

Index

About the Author

Nina Beth Cardin is a community rabbi who works to promote environmental health and environmental justice. She is the Chair of the Sustainability Sub-Committee of the Social Justice Commission of the Masorti Movement and the co-author (with her husband, Rabbi Avram Israel Reisner) of the Conservative Movement's teshuvah on sustainability. She has worked in the field of environmental advocacy for over twenty years, most recently promoting environmental human rights and intergenerational environmental equity for all.

She was in the first class of women ordained by the Conservative/Masorti movement. As a community rabbi, she has founded several organizations including the Jewish Women's Resource Center (NYC), the Pregnancy Loss Support Program, the National Jewish Healing Center, the Baltimore Orchard Project, and the Baltimore Environmental Sustainability Network. She has spent the last few years advocating for the constitutional protection of environmental human rights.

She is the author of numerous articles and several books, including *The Tapestry of Jewish Time: A Spiritual Guide to Holidays and Life-Cycle Events*; *Tears of Sorrow, Seed of Hope: A Jewish Spiritual Companion for Infertility and Pregnancy Loss*; and *Out of the Depths I Call to You: A Book of Prayers for the Married Jewish Woman*.

She lives in Baltimore, Maryland.